The Art of Science

A Practical Guide
to Experiments,
Observations, and
Handling Data

Joseph J. Carr

HighText
publications inc.
San Diego

Library of Congress Cataloging-in-Publication Data
Carr, Joseph J.
    The art of science : A practical guide to experiments, observations, and
handling data / Joseph Carr.
        p. cm.
    Includes bibliographical references and index.
    ISBN 1-878707-05-1 (pbk.) : $19.95
    1. Science—Laboratory manuals. 2. Science—Methodology—Handbooks,
manuals, etc. 3. Science—Experiments—Methodology—Handbooks, manuals, etc.
I. Title
Q182.5.C37 1992
507.8—dc20   92-70471
                                                      CIP

Printed in the United States of America

Cover design and illustrations: Brian McMurdo, Ventana Studio, Valley Center, CA
Developmental editing: Carol Lewis, HighText Publications
Production services: Greg Calvert, Artifax Corporation, San Diego, CA

ISBN 1-878707-05-1
Library of Congress catalog number: 92–070471

HighText is a trademark of HighText Publications, Inc.

7128 Miramar Road, Suite 15
San Diego, CA 92121

Contents

Preface

Science is a verb. Although grammar experts may disagree, it is nonetheless true that science is a field for *doers*. This book is intended as a survival manual for newcomers to science who are entering the "doing" phase of their careers. It will show you how to do good experimentation, and how to avoid some of the pitfalls on the pathway—and there are many. It is written to accommodate a wide range of levels from senior high school to graduate school, and a wide range of different interests including the physical sciences (e.g., chemistry, physics), life sciences (biology, physiology, and so forth), and social sciences (psychology, sociology, and the like).

My years in a medical research environment, in graduate school, and later on judging science fairs and refereeing scientific and engineering technical papers and book manuscripts have led me to believe that there is a tremendous lack of understanding of how the research *process* works. In addition, there is a general misunderstanding of the principles of arithmetic used to process data derived from scientific experiments or field observations. Indeed, one of the most common reasons for the failure of graduate school theses and dissertations, undergraduate senior papers, and professional publications is the lack of proper statistical treatment of the data. Therefore, part of this book discusses the elements of statistics and other topics in the numerical treatment of data.

Also discussed in the book is material on the presentation of data, both in report and graphical form. Reports are often improperly drawn, and as a result readers either fail to draw as much information from the report as would otherwise be possible, or manage to find much fault with the effort simply because of shortcomings in the reporting method. Similarly, graphs are often misdrawn, misconstrued, or even used to support lies. We will discuss the use and abuse of graphs in Chapter 13.

We will also take a look at critical thinking, a subject of vast importance to science. Chapter 2, entitled *Thinking Scientifically* deals in depth with various problems that creep into scientific works, as well as into everyday life. Understanding these problems will help you avoid making them yourself and recognize them when others make them.

Finally, for the youngest readers, or for those teachers who are mentoring them, Chapter 15 tells how to win a science fair. It is surprising how often little things cause a loss of first— or any—place in science competitions. A lack of attention to certain details can be devastating.

Whatever your scientific field of interest, or whether your goal is amateur science or a profession in science, this book is designed as your standardized, generic, plain-vanilla *survival manual*...Keep it handy.

Joseph J. Carr

What is Science About?

HE WORD "science" denotes a wide variety of activities that are so different from each other that their respective practitioners seem to speak different languages. While the various jargons spoken by field biologists, physicists, and sociologists are vastly different, there is one factor that links all scientific fields together—reliance on investigative methods that force us to look at the world in a systematic and rational way.

Some people break science into the physical sciences (e.g., physics, chemistry), life sciences (e.g., physiology, botany, molecular biology) and social sciences (e.g., sociology, anthropology, psychology). Some assert that the physical and life sciences are so-called "hard sciences," while social sciences are "soft sciences." Other people imply that the social sciences are not true sciences at all, because their fields of study often involve factors that are not, so it seems, strictly physical. For example, a sociologist often deals with irrational and self-destructive customs or behavior in one population or another. Perhaps such opinions result from an overly strict definition of the word science. Or perhaps the word science is used for those fields of study because our language proposes no other word that defines the *systematic* study of phenomena. For our present purposes, the word science is taken to mean all of the scientific fields of study, provided a systematic approach is taken to study the phenomena. Indeed, dictionaries often list the strictest

definition of science last, or next to last, in the order of preferred definitions. So, we may state that...

> *Science is that endeavor that seeks to study, or obtain knowledge*
> *about, phenomena through the use of a systematic approach*
> *that is based on the evaluation of evidence through reason.*

Misconceptions

To the layperson, science seems like a precise thing with
immutable laws that forever govern the way the universe
works. Such may ultimately prove to be the case, but science
is still an imprecise field, full of ambiguities and half-truths,
because of the still imperfect state of our knowledge. If this
weren't so, then there would be nothing left to research—
we would know it all!

Because of the imperfect nature of our current scientific
knowledge, we often find old concepts being overturned and
new concepts emerging in their place. Favorite theories are
raised up and thrown down in short periods of time. To be
engaged in science, you must be flexible, be capable of handling
uncertainty, must view all knowledge not as absolute but rather
as merely tentative—no matter how well that knowledge
seems to be proven—and must be quite willing and able to
change your mind as new evidence calls old theories into
question. False pride of authorship of a pet theory that is no
longer supportable has ruined more than a few scientific
reputations.

An example of a massive change in scientific thought,
one that caused many scientists long periods of severe doubt,
occurred in the first three decades of the twentieth century.
For nearly 300 years, until about 1900, science viewed the
universe in terms of Isaac Newton's laws. Things in the classical Newtonian universe followed nice, orderly, mathematically
correct paths according to fixed, easily understood rules. So firm
were those rules that they were (and still are!) called *laws*. It

can be argued that the reputation of science for bedrock certitude rests on the spectacular success of Newtonian physics—not to mention the arrogance of some scientists of the time who were partial to the idea of their own near infallibility.

But late in the nineteenth century, phenomena began to be noticed that did not fit nicely into the classical Newtonian framework. By the end of the first decade of the 1900s, these problems were solved as new ideas emerged that shook the very foundations of the old physics. Albert Einstein proposed the Theory of Relativity, and physicists such as Niels Bohr, Erwin Schrödinger, and Werner Heisenberg developed quantum mechanics (QM) to explain the atomic world. But even these giants didn't always agree, no matter how compelling the arguments. Einstein, for example, never quite accepted the ambiguities inherent in QM, disdainfully claiming that "God does not play dice with the universe."

The goal of this discussion is to bring you to a point where you understand fully that uncertainty will dog your scientific career... and that you should view that situation with eagerness and excitement! Hopefully, as each study progresses, uncertainty will drop—for awhile—only to be replaced with new uncertainty as newer research and study causes new questions to suggest themselves.

The language of science is numbers—that is, mathematics—and the particular area of mathematics most useful for dealing with uncertainty is statistics. Until statistics began to be applied to science there was much wasted effort. In 1935, Englishman Sir Ronald Fisher said: "The waste of scientific resources in futile experimentation has, in the past, been immense." Fisher's 1936 book, *The Design of Experiments*, formalized the application of statistical knowledge *before* starting an experiment; he also formalized the field of "analysis of variance." Some of the material in this book is a primer on statistics and numerical data. But it is only a very basic introduction, and does not take the place of a proper course in statistics. Russell Langley's book, entitled *Practical Statistics*, is an excellent reference. (See the Chapter 7 References.)

Approaches to Doing Science

Science requires a systematic approach to gathering knowledge about a subject. What constitutes "systematic" depends on the specifics of the phenomenon being investigated. There are, however, three general types of investigation: (1) ordered (or informed) observations, (2) surveys, and (3) directed experimentation. Which approach or combination of approaches is used depends on the nature of the scientific knowledge being sought.

Ordered/Informed Observation

This method is the province (mostly) of field biologists, geologists, and others who must make observations under less than perfectly controlled circumstances. But the fact that you are dependent on the coincidences of having the right species of animal wander by, or finding a particular plant in the woods or the right rock formation, does not alleviate the need for ordered working habits. After all, "coincidences" have a peculiar way of favoring the well prepared.

The field scientist goes to make his or her observations by having a plan in mind of *what* to collect, where it should be sought, how it should be systematically recorded, and how observations should be interpreted. In nearly all cases, the researcher will try to make the observations in such a manner that the investigations do not interfere with what is being observed. If you want to learn how crows behave, it is best to conceal yourself so as to not alter crow behavior by your mere presence.

Surveys

A survey is also an investigation into "... things as they are, without interference" (as defined by Russell Langley in *Practical Statistics*). But, unlike ordered observations, surveys may involve some interaction between the surveyor and the subjects being studied. Social scientists often use surveys in order to get information that may otherwise be unavailable under any other circumstances.

Langley's book defines three basic forms of survey: retrospective, current, and prospective. A retrospective survey reaches back into the past to find out how things were. This form of survey is often severely hampered by several important factors. First and foremost is that data cannot always be arranged in as pure a form as needed to reduce (or hopefully eliminate) confounding co-factors. Second, the data may be of the correct form, but may be incomplete or even missing. The further back one reaches, the less reliable the data becomes. Not only did we (or our ancestors!) not know enough to retain the specific data required, but we may not have even had any interest in collecting the data in the first place!

A *current survey* looks at things the way they are now. You may seek a profile of weather conditions in your state or region on a specific day or week. Or, you may attempt to find out the attitudes of the American population on a specific issue or practice.

A *prospective survey* selects a population today and then follows it for a period of time into the future to see what changes. An example is the tumor registries maintained by cancer researchers. Another example is the Framingham (MA) study of heart disease that followed a specific local population for a number of years.

Experiments

Again quoting Langley, an experiment is a contrived event in "... which the effect of some deliberate act is observed." In

other words, the investigator sets up a situation in which a phenomenon can be observed to change so that confounding factors are either well controlled or eliminated altogether. For example, simple random observation may show that a bright flash of light and some water are created when hydrogen and oxygen gases are ignited by an electrical spark. But that observation, no matter how valid it turns out to be, is merely anecdotal—i.e., a good "war story"—unless you contrive an experiment where hydrogen and oxygen gases are placed into an evacuated vessel, making sure that only H and O_2 are present, and then introduce a clearly defined electrical spark into the vessel. . . followed by an evaluation of what you observed. Either type of observation, chance or contrived, is worthy of being recorded in a scientist's notebook, but only the contrived experiment helps to establish understanding. The chance observation is useful as a starting point for asking a question of nature. . . but it only suggests a pregnant area of research, and doesn't usually define it.

Some people think that science is only common sense that is mathematically validated. Let's see about that in the next chapter. . . you'll find that an all-too-*uncommon* sense is the key to good science!

2

Thinking Scientifically

GOOD SCIENCE requires good thinking—the two are inseparable. Good thinking is a purposeful activity. It helps to think a little about thinking from time to time. In this chapter we will look at different modes of thinking and discuss which ones are appropriate to scientific activities.

In his book *Critical Thinking*, Richard Paul divides thinking into two general categories: *monological* and *multilogical*. The monological problem has ". . . only one correct answer, with a single or convergent logic of justification for it." Paul cites physics problems at the end of textbook chapters as good examples of monological problems.

Such problems usually succumb to procedures for solution. That is, they are said to be *algorithmic*: If you follow the right procedure, then you will necessarily come up with the right answer.

A multilogical problem, on the other hand, may have no truly correct answer. Or, the answer may be dependent on the point of view of the problem solver. Such problems are routinely seen in psychology and social science. It may also be that the field of inquiry is so new that old ways of thinking don't come to grips with the new material. Old definitions become less useful, old procedures yield wrong answers, and truth seems trapped in the wrong frame of reference.

Monological thinking is appropriate when a field of knowledge is well developed and procedures can be defined for solving the problems. Monological thought is not simply the province of students and those who teach them, however. It also works in research that extends knowledge of a field. But monological thought becomes less and less useful as the field becomes more and more ambiguous. When things are not so well known, or when totally new territory is being explored, then multilogical thought works best.

Some scientists sneer at multilogical thought. Such an opinion, however, reflects someone who is perhaps well schooled in science but who has an unfortunately narrow view of things. Breakthrough scientific discoveries tend to be made by those who have the integrity to challenge older views of reality, and that often requires a multilogical approach. Einstein would surely have been drummed out of college if he had advanced some of his best ideas as a student. The person who holds a negative attitude towards the multilogical may be well versed in solving scientific problems of the past, but may have difficulty making anything more than minor advances in an older field. New fields, or new lines of study, may be difficult for them to comprehend. When something new is encountered—my favorite example is the crisis of the Newtonian physicists between 1900 and 1930 when scientists came face to face with quantum mechanics—they can't cope because they are only comfortable thinking in the boxes previously defined for them by others.

People of true vision are able to "shift gears" back and forth between the monological and multilogical modes of thinking, while others are stuck in neutral (or even reverse).

Reductionism and Holism

Thinking is the proper approach to solving problems, regardless of whether they are old or new. Here again we can divide the subject into two categories. (The world is divided into two kinds of people: those who divide everything into two categories, and those who don't!) First, there is the *reductionist* approach and, second, the *holistic* approach. I am a little nervous using the word "holistic" because of the connotation of flaky food or health faddism or mystical New Age vacating of all thought, but it does describe the concept well. Perhaps "integrative thinking" is a better term. The idea behind holism is to synthesize new knowledge by looking at a field as a whole and by integrating different fields of knowledge.

The reductionist way of thinking is the means by which science progressed for most of its modern history from Copernicus and Galileo to the present time. It is powerful and leads to progress— even breakthrough progress—when done properly, so it is how science will progress in the future. In the reductionist mode of thinking, a problem is broken down into little pieces, and each piece is then solved separately. A metaphor used by one teacher was the chore of eating a steak. If you attempt to eat the entire 12-oz. steak in one bite, then you will very likely choke to death. But if you cut it into 15 or 20 small pieces, then you can chew each piece individually and eventually accomplish the job of eating the steak without fatally blocking your trachea.

The principal advantage of reductionism is that it permits us to deal with smaller problems that we can solve, rather than unsolvable massive problems. Supposedly, when we reintegrate the fractured subject area-knowledge gained through this process, we get a good picture of the whole. But there are several problems with reductionism.

First, we may provide ourselves with a faulty subdivision of the entire field in the first place. There are sometimes holes in

"Yeah, it'll work"

the coverage of a field. And those holes are often generated by the process of what makes sense to older scientists who may or may not be willing to invest time in new ideas that upset old paradigms. Research projects tend to cover what graduate students' advisors think is appropriate and what can generate grant money. It is generally a very good process, but not without fault... so be careful.

The reintegration of the smaller pieces may be faulty, or may be lacking altogether. If no one has a "big picture" view of things, a serious lack of insight will hinder pulling the fragments together into a whole. "Synergy" is an interesting word to contemplate in this context.

In addition to spotty coverage of a field, reductionism may create a "think-in-boxes" mentality that causes people to be less aware of developments in adjacent areas of research or other fields. As a result, new advances are stymied or delayed because of miscommunication through over-specialization. It is tragic to see situations like Edward Lorenz's work in chaos being overlooked for many years because it was published in a weather science journal that other scientists rarely read. Indeed, the related fields of fractals and chaos are filled with such miscommunication between the various branches of science.

A serious concern for purely reductionist thinkers is the possibility of committing the fallacy of wrongful division (covered in Appendix A) or the possibility that the reduced fraction may not truly represent the whole. Without a bit of holistic integration, for instance, it is hard to see how any amount of knowledge about the cells of the heart lead to a clinical knowledge of heart disease.

The holistic method is one that looks across disciplines and attempts to integrate knowledge from a lot of different fields. It is not merely some ambiguous idea, but rather an active attempt at synthesis of new knowledge from the fragments gleaned from various fields. It looks for patterns and similarities between fields that have completely different jargons to describe like things. It will consider the possibility—

horror of all horrors— that it is on the wrong track. Or that another scientist's ideas on the subject are nearer the truth than one's own. Holism takes hard work, and it can be painful, but it is worth the trouble.

It is the job of schools to perform the knowledge integration function, but many are seriously derelict in this duty. School courses tend to be as narrowly divided as the professions that originate the knowledge, and as a result students rarely integrate two or more disciplines. Of course, there are master teachers, magnificent mentors, and others who perform this function, and they are to be dearly cherished. In my opinion, Richard Paul's book *Critical Thinking* should be required reading for all educators... and is highly recommended for everyone else who professes to be educated. (See the chapter references.)

Inductive and Deductive Thinking

Approaches to thinking are sometimes divided into *deductive* and *inductive* classes. Both are needed in science, although some would assert that only the deductive form qualifies.

Inductive thought examines a collection of evidence, and then proposes a theory to account for all of the observed facts. The inductive thought process starts with the specific and progresses towards the general.

Deductive thought processes are the inverse of the inductive. A hypothesis is proposed, and then data is collected to confirm or disconfirm the hypothesis. Deductive thought proceeds from the general to the specific.

Scientists use both inductive and deductive processes in their work. Whether inductive or deductive is used depends on the situation and the field of study. There is often an iterative situation in which both induction and deduction are used alternately. Observations of nature are made, and theories are inductively formed to explain the observations. Then, predictions are made and additional confirming or disconfirming evidence is sought through deductive processes.

Lateral and Vertical Thinking

The lateral and vertical division in thinking modes was proposed by deBono in 1968. Vertical thinking bores down on a narrow aspect of a problem and probes deeply to find a solution. Lateral thinking is broader viewed, and may "look around" the problem, rather than at it. Lateral thinking tends to be creative and is engaged in idea formation and discovery. Vertical thinking, on the other hand, concerns itself with refining and developing ideas. In science, lateral thinking is useful for generating new theories, novel ideas, and advanced concepts, but vertical thinking is required to develop them and refine them to a point where general acceptance is possible. The individual who is capable of both forms of thinking is a "power person"!

Common Sense and Science

It is commonly believed that science and common sense are closely connected. Before we accept such a statement, however, it may pay to examine just what is meant by "common sense" and see if it applies to science and technology.

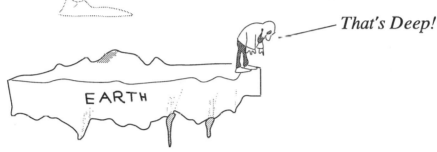

That's Deep!

The common-sense viewpoint?

Common sense (also called "horse sense") is one of those concepts that nearly everyone supposedly understands and accepts intuitively. Common sense is widely held to be a virtue—perhaps the chief virtue—in business, professional, and scientific settings. A person who routinely makes prudent choices and gives wise advice, even when lacking in formal education or scholarly credentials, is said to possess the attribute called "common sense." On the other hand, a person who lacks common sense can't quite seem to do anything right. (As a Southern humorist I heard on the radio said, they are the kind of good ol' boys who would dearly love to shoot straight, but can't quite find the trigger.) Paradoxically, if a person exhibits a large measure of all-too-uncommon *good* sense, then we impute to her a large measure of "common sense."

As so often happens, universally held "truths" turn out to be false when closely examined. When the concept of common sense is systematically examined, it turns out that it's not really a virtue at all. It may well be that common sense is a serious impediment to progress in general and problem solving in particular. The dictionary is usually a good place to start when examining the meaning of words. For common sense, one of Webster's definitions is "... the *unreflective* opinions of ordinary [people]."

The Roots of Common Sense

Notions about what constitutes common sense in any particular context are found in the world views of the people who hold those notions. A world view is, to use a much abused word, a *paradigm*, that is, a thought pattern through which a person filters all perceptions. The paradigm built by each individual is the unseen collage of nonconscious ideologies, unstated assumptions, unexamined opinions, and automatic habits of thinking that we all bring with us by simple virtue of our backgrounds. Everyone has a world view, and everyone's world view is slightly different from others. And while such diversity can lead to wasteful conflict, it is also a source of profound strength. After all, the creative tension caused by mixing differing viewpoints can lead to insights that might otherwise remain hidden.

How does this perception screening affect scientific problem solving? If common sense is the filter that screens

perceptions, then it will limit the allowed solution possibilities to the familiar—i.e., to what worked in the past. As a result, old assumptions are not tested, old methods are not revalidated for new situations, and new problems are force-fit into old forms of solution. We tend

to see new problems, and their potential solutions, in light of
old realities as if those realities are timeless, immutable
truths. That was the principal error of Newtonian physicists
around the turn of the century who were greatly troubled by
relativity and quantum mechanics.

Why is it important to see things in a different light? The
reason is simple: *the world today—especially science and technology —is changing at an unprecedented rate.* If we need to adopt a
symbol for our age, then let it be the Greek letter Δ (*delta*)—
the symbol used by scientists to indicate change. Old solutions
may or may not work in any given situation today. The one
thing that is certain is that a rapidly changing world is both
challenging old paradigms and creating new ones to meet new
realities.

The chief problem with common sense is that its limited
focus makes it unable to deal with those rapidly changing
realities, especially when the old predictors of future behavior
of the system fail us. In other words, common sense solutions
are self-limited.

Another failure of common sense is that the underlying
assumptions of our world view are not questioned. Unless such
assumptions are occasionally re-examined, they will eventually

not only become less valid, but can have a terribly damaging effect.

Common-sense approaches to problem solving frequently fail because they proceed from an uninformed basis. Evidence may be lacking, or missing, or entirely misconstrued. More is sometimes made of available evidence than is warranted because it is filtered by unspoken assumptions, so seems to need no additional proof. I'm reminded of a little scene that reportedly occurred at a fancy downtown luncheon restaurant. A business executive finished her meal, paid the check, and then went to the cloak room for her coat. She was surprised to find the headwaiter already standing there with the correct coat in his hand, ready to help her with it. "Amazing! How did you know that was my coat?" said the guest. "I don't know that it's *your* coat, Madame, but I do know that it's the one you came in with. . . . " Scientists cannot afford to make more of what they know than is justified by the available evidence.

Still another problem with common sense is that it tempts us to take action without really understanding the problem. At a Navy management lecture, a young Lieutenant Commander objected that ". . . this stuff is all well and good but. . . ," he averred [*pregnant pause*], ". . . the Navy pays me to make decisions." His remarks imply that good decisions are fast decisions, and indeed in dire military emergencies "quick draw" decisions are often absolutely essential. But not all decision making is best done fast. Perhaps what the young officer did not understand is that the skills needed to react to an incoming surface-skimming missile are not necessarily the same skills that will yield the best results when finding the problem with a faulty industrial process. . . or in doing research. One might be tempted to refer the "quick draw" crowd to Warren Bennis's book on leadership: "It is not enough. . . to do things right; [you] must *do the right things*." (If you're interested, you'll find this book listed in the references at the end of this chapter.)

The Opposite of Common Sense

The alternatives to common sense are not, as some would suppose, either brainless foolishness or vain "pie-in-the-sky" musings. The opposite of common sense is *good sense* informed by knowledge gained through systematic examination of the problem to find its actual cause. Once the cause is found, you can then invent and test possible solutions (experiment) and then implement a change that corrects the problem.

The basis for scientific knowledge is information based on valid data. Keep in mind the differences between these three concepts: *Knowledge* implies understanding of the problem or process being considered. (Knowledge is created by correctly interpreting the information at our disposal.) The key to the word "information" is the *inform* part—information must inform the receiver about the problem at hand. *Data* are merely facts devoid of context. Not all data are valuable: The right data must be collected, with accuracy and precision, and then placed into the proper context for the problem at hand. Context gives meaning to data that converts it into information. We can therefore say that information is formed from data imbued with context in such a way that it becomes knowledge. Scientific advances are based on knowledge, not "gut feelings" or common sense. Knowledge is gained through the process of systematic research.

Systematic Research

The key to good systematic research is the Confession of Ignorance: "I don't know." (Perhaps children's pet phrase "I dunno" means that they are research geniuses!) You first have to realize that a deficiency in knowledge exists, and then resolve to correct the situation. When trying to discover something about nature, you must *ask a question of nature*, and then create an experiment or data collection plan that can answer that question "yes" or "no." The question should be narrowly and clearly formulated. "Global" questions tend to be so difficult to pin down that the effort is likely to fail in the end.

It's not necessary to make the question profoundly deep, as no one expects you to break totally new ground out on the furthest frontiers of knowledge. (It could happen, however!) Indeed, focusing on stellar profundities may be satisfying, but may also cause you to overlook the actual problem!

What Is Critical Thinking?

Critical thinking is an approach to problem-solving and decision-making. Good science can't be done without it. The basic processes of critical thinking include planning, information gathering, defining a frame of reference or context, monitoring, accounting for biases, and evaluating.

These processes must be applied from the very beginning. If we misconstrue a problem to begin with, then the solution is also necessarily in error. Computerniks reduce this concept to a single terse acronym: GIGO (garbage in, garbage out).

Critical thinking involves recognizing and accounting for the assumptions, background logic, biases, intentional deception, and other factors that tend to distort the outcome of scientific thought processes. Truly objective observations are difficult to obtain, especially in the social sciences, because all people are encumbered by the world view that I've previously mentioned—their unique set of personal values, hidden assumptions, presuppositions, and opinions that colors everything they think or do.

Sometimes there is also a special background logic in place. For scientists, this background logic is formed from their prior education, culture, and personal ideas about what is true and what is false. When something new comes along to challenge that background logic, it is often seen as aberrant or just plain wrong. With proper method, the adverse effects of background logic can be minimized. It is the role of the critical thinker to probe into their own, and others', background logic when a claim to truth is asserted.

In order to advance the state of knowledge in any field it is often necessary to think the unthinkable and find out where it leads. Copernicus and Einstein are brilliant examples. But so is the pharmacologist or botanist who takes primitive healers seriously and discovers a potentially helpful medicine in their rude herbal concoctions. The critical thinker has the ability to put himself or herself into the opposing point of view, not in order to refute it but rather to see if it is true.

The ideal scientist is a critical thinker who is humble enough to make the Confession of Ignorance, self-confident enough to admit it out loud, and courageous enough to set out on a course that will rectify the situation. . . even when others think there is no sense in it. And courage is what it sometimes takes. I recall a physicist who made observations that contradicted a theory for which his department chairman was justly famous. The theory was well regarded in its time, but later information proved it to be incomplete in some areas and wrong in others. My friend told me that he'd known it for years, but was afraid to speak up because of the personality of the chairman (although he held precious little *real* power over other research faculty, it turned out). Why the silence? Perhaps it's because "silence isn't always golden, sometimes it's just plain yellow." (Frank Eiklor said this on *Shalom Radio*, in response to a question on why he fights anti-semitism in the face of threats from neo-Nazi elements.) To be a great critical thinker, you must be willing to announce, during your Nobel acceptance speech, that the research for which science's greatest prize was just conferred on you was all wrong. . . that later data showed a different conclusion.

Critical thinking is of great importance to all educated people, yet it is a rare and precious thing. According to a 1983 study of the National Commission on Excellence in Education the United States is a "nation at risk" because of a profound lack of instruction that trains students in the ability to think. While knowledge, information, and data proliferate at an exciting and breathtaking rate, the amount of time spent by schools in training students in logical abstract thought is diminishing.

Although most college degree programs once included courses designed to teach thinking, and many other courses consciously incorporated critical thinking skills practice, more and more colleges are herding their students towards knowledge-based courses that lead to jobs in the near term. "Trade school" universities have recently been touted as the universities of the future. But such vocational college education is really of limited value because it cheats the student out of long-term benefits. I've heard it said that true education is what survives when a person has forgotten all that he has been taught. In a dynamic culture where technical and scientific knowledge is raised up and overthrown a dozen times in the course of a scientist's career, the ability to think clearly is of far greater importance than the ability to regurgitate facts and formulas that are of value today. . . but obsolete tomorrow.

Scientists and engineers often mistakenly believe that they are well trained in critical thinking because their education and usual career path involve substantial exercises in problem solving. But the analytic type of problem solving learned by scientists and engineers is not the same as critical thinking. Most of the problems faced by such students, for example, are strictly monological. Such problems are inherently easier to solve than multilogical original research problems. Furthermore, once you get away from the types of problems that succumb to a differential equation, a bit of algebra, or a numerical analysis, our methods fall down and may actually burden us. Abstract thought, and solving problems for which there is neither a correct answer nor even a good answer, is considerably more difficult.

Brainstorming

All of us occasionally fall into deep intellectual ruts and seem unable to climb out. In other cases, one or more blocks to thinking occur, forcing a loss of time and energy. In still other cases, solutions that later appear to be so simple and straightforward never occur to a particular individual researcher. But there is a way out. *Brainstorming* uses the synergistic effect that takes place when a number of people get together and contribute in an informal, nonthreatening environment where people are free to express "silly" ideas. Another term for this activity is *supersum thinking*.

In a brainstorming session, all possibilities and impossibilities that participants think up are laid out. One prime rule: No criticism or ridiculing of "bad" ideas, no personal "zingers," and no personal references to the participants or their respective abilities are permitted. Once a large number of possibilities are on the table, then a distillation process can take place.

A strange and beautiful thing can happen in a brainstorming session. As possibilities narrow, more and more brilliant variations surface. Eventually, if the process works properly, a certain synergy is reached, and a final product is produced that is truly "greater than the sum of its parts."

Formal brainstorming can be taught to people. But the formal methods are not always needed because elements of it can be adapted for specific situations. Sometimes all that is needed to make the thing happen is a nonthreatening, collegial atmosphere coupled with a willingness to consider the supposedly absurd. "Coffee-shop" or "wine-and-cheese-party" environments often produce the needed atmosphere.

A principal danger in brainstorming is getting off the track into personalities or egos. If anyone is allowed by the group to attack ideas as "idiotic" (even though they may *be* idiotic), then a chilling effect sets in and dampens the whole process. Given the right negative factors, a brain storm can suddenly degenerate into a brain drain.

Blocks to Good Thinking

Even the best of us from time to time fall into several different kinds of intellectual traps that hamper good thinking. Some of these are the formal logical fallacies, or false arguments. False arguments are not always easily spotted. Sometimes, the falsehood of an argument is subtle and isn't quickly recognized. At other times, an argument is so blatantly false that anyone could see the error. In still other situations, paying too little attention to an argument as it is presented may lead you to a premature acceptance of it (especially if it initially seems valid, but on closer examination is revealed to be false). It is all too easy to fall into that trap, especially when you are not alert.

In the hands of an honorable person, reasoned argument is a very powerful tool. But in the hands of a fool, a dishonest person, a propagandist, an incompetent, or any other person who lacks either the skill or the integrity to persuade properly, then an argument can either deteriorate into fallaciousness, dissolve of its own contraditions, or be perverted with necessarily bad—sometimes evil—results.

Not all fallacies result from fraudulent intent. Sometimes they result from a weak position created by either bad information, faulty reasoning, or sloppy thinking. The scientist may be perplexed as to why his or her theories are not accepted, why the argument is not a winning one, or why some other theory was, in the end, accepted.

It is often the case that the person attempting to convert you to their position uses one or more fallacious arguments in their presentation. In either case, whether you are the persuader or persuadee, the best defense against faulty argument is the ability to recognize and deal with the basic types of fallacy. Whether a faulty argument results from honest error, polluted data, or from fraudulent intent, the result for you is the same: acceptance of error. By learning the various types of fallacies, and being able to recognize them, you can guard against such problems. Such skills can only increase your scientific competence. I've included an unofficial list of the various types of logical fallacies in Appendix A.

There are a number of other intellectual traps that scientists must also seek to avoid.

Failure to seek disconfirming evidence

When initial data tends to confirm a predetermined or inherently comfortable position, there may be a tendency to overlook disconfirming evidence. A relatively common example is the scientist whose reputation is based on a certain finding. The scientist with the vested interest may collect data that tends to confirm his or her idea, while overlooking the data that either confirms an alternative proposition or disconfirms the pet theory.

Wishful Thinking

This block to good thinking is also known as the Pollyanna Principle, after the early twentieth century novel about a young lady who refused to see anything wrong. There are different ways that the Pollyanna Principle can strike. In one type of situation, the person is so convinced of a positive outcome that he overlooks reality. For example, when a person buys a state lottery ticket, especially when the prize is a whopping sum of money (at least in annuity value), they may delude themselves into thinking that it is already won. . . despite real odds of 1 in 7,000,000. Nobel prizes may also be seen that way (*sigh*). On a more mundane level, you may wish so fervently for a positive outcome to an experiment that you overlook critical factors or important contrary data. A Pollyanna attitude may be cute in a little child but has no place in the hard-headed reality of science and technology.

Entrapment

"We've already invested so much time, effort, and money in this research project that we've simply got to finish it!" typifies this problem. There comes a time when a project, a theory, or a pet belief is simply untenable, and no amount of money or effort will make it right. If the desire to succeed in a research project is so strong that one overlooks evidence that it is a "no-go," then one has lost an even larger game. Scientists and engineers cannot afford emotional attachment to causes so strong that they cannot see the reality of a lost situation, especially when research budgets are shrinking.

It is disheartening, on the other hand, to see good ideas squelched for the lack of resources to conduct the research. While you must guard against the resources-entrapment pitfall, you must also be sensitive to the possibility that a really good idea merely got off track, and could be salvaged with only a little more effort. A truly interesting dilemma—especially out on the frontiers of human knowledge.

Ego Traps

The human ego is very powerful, and even critical thinkers can sometimes fall into this trap. I saw an example once in a medical center research project. A cardiology project was one of the first groups to successfully diagnose human electrocardiograms (ECG) by computer (a task that is common today, but considered novel at the time). The computer programmer configured the program so that the analog readout of the digitized waveform stayed on the screen during the time when the computer was performing a binary search tree routine on a hard disk file of more than 25,000 confirmed pathology ECG waveforms (a simple but lengthy pattern-matching task). The search usually took a couple of minutes, most of which was spent waiting for the remote computer to communicate. While they were waiting, however, the emergency room doctors would examine the analog ECG waveform on the screen and form their own opinion on the diagnosis. When the computer reported a different diagnosis, the doctors universally assumed that their own diagnosis was correct, and the computer was wrong. Retrospective studies by a board of five qualified electrocardiographers, however, showed that the computer was right about 85 percent of the time, while the ER doctors were right only 40 to 60 percent of the time.

Incidentally, the programmer eventually moved the analog readout to the end of the process, after the proposed computer diagnosis was on the video screen and the printer was stitching a hardcopy printout. When the analog waveform was not made available until after the diagnosis was on the screen, the docs didn't form any preconceived opinions. That move was the only change to the computer program, but the ER doctors were thrilled at how much he'd improved the diagnosis algorithm!

Summing It Up

Should the scientist be monological or multilogical? Should the scientist be reductionist or holist? Should the scientist be inductivist or deductivist? lateralist or verticalist? In a word: *Yes.* As a thinking person, the scientist needs to be able to draw on all of these thinking skills, or risk being little more than a technician with a lab coat. Perhaps the best metaphor is the zoom lenses used on cameras. You need a Zoom Mind that can change focal length from wide angle to long distance with, of course, a high degree of close-up macro capability (and equipped with special filters for discerning intellectual traps).

References

Warren Bennis, *On Becoming a Leader*, Addison-Wesley (Reading, MA, 1989).

Diane F. Halpern, *Thought and Knowledge: An Introduction To Critical Thinking — 2nd Edition*, Lawrence Erlbaum Associates (Hillsdale, NJ, 1989).

Richard Paul, *Critical Thinking: What Every Person Needs to Survive in a Rapidly Changing World* (A.J.A. Binker, editor), Center for Critical Thinking and Moral Critique, Sonoma State University (Rohnert Park, CA, 1990).

Claire Selltiz, Lawrence S. Wrightsman, and Stuart W. Cook, *Research Methods in Social Relations—Third Edition*; Holt, Rinehart and Winston (New York, 1976).

Theory, Hypothesis, and Law—What's the Difference?

3

THE WORDS *theory* and *law* are used quite a bit in science, but they are often misunderstood. Many people, when they hear the phrase "scientific law," immediately think of something that is both universal and unchangeable. But in science a law is merely an observed regularity in nature. . . or a statement of order or the relation of phenomena that *so far as is known* is invariable *under the given conditions.* Thus, a scientific law is neither universal nor immutable, but rather it is something that is always observed to happen the same way—at least that is the way it was in the past.

The use of the word "law" does not in any way exclude the possibility that someday, under some circumstances, we might find an exceptional observation that forces us to change the law. It sometimes happens that refinements in measuring techniques, improved experimental apparatus, or new observations that were previously unsuspected lead us to reconsider the concepts underlying some specific scientific law. Like the legislative kind, scientific laws are subject to change. A good example is Newton's laws. For many generations Newton's statements about the physical universe remained unchallenged because there were no observations that either were not explainable by Newton's laws or that contradicted Newton's laws. However, by the turn of the twentieth century, observations in thermo-dynamics and other fields of science were difficult to fit into the framework of Newtonian physics.

It must have been both a heady and disquieting time for scientists: Breakthrough leaps in knowledge, rising public perceptions of the infallibility of science, breathtaking advances that continue today, all coupled with a mounting horror over a few observations that could not be easily explained by experimental error or measurement "gremlins"; there were real, daunting problems in the observed data.

By the simple idea of assuming that energy is quantized, Max Planck in October 1900 pulled on the first thread that unraveled the fabric of the Newtonian construct. Poor Planck, being thoroughly a classical Newtonian, spent the rest of his life not quite believing the ideas that fell out from his own idea of energy quanta. We now know, after Einstein, Bohr, Heisenberg, Schrödinger and many, many others thought and did science, that Newton's laws were never immutably true, but apply only to the narrow range of circumstances in which we live our ordinary lives. At velocities close to the speed of light and magnitudes on the galactic order, Einstein's relativity is a better (but probably imperfect) model; at sizes on the order of the atomic, quantum mechanics is the better model. But, as our everyday experience attests, Newton is alive and well in the world in which we live.

Any scientific law is merely a *tentative* statement, or intellectual construct, of observed invariability of certain phenomena or their interrelationships. Your scientific career will be quite exciting if you routinely expect to see laws modified, or cast aside altogether, in the course of your lifetime. I suspect that a Nobel Prize or two is lost from time to time by people looking for the expected, and thereby overlooking or discarding anomalies that don't fit the preconceived mold. Was a peculiar data point experimental error or was it nature whispering a hint of a secret in your ear?

Science may well be too dull to contemplate if we ever really do find the elusive Theory of Everything. Some scientists are talking about writing the final chapter in physics, or finally being able to "read the mind of God" (as one put it), but I'm convinced that there is as much yet to learn as there is known, and that's the excitement of it all. I suspect that any proponent of a closed book on science is really a sad soul admitting to being burned out and no longer having the vision to see anything left worthy of getting excited about.

Theory

Some people reduce theory to the status of uninformed opinion, of which all possible variations are equally acceptable or disdained—except, of course, one's favorite theory which may look like a rock-solid fact. Other people look on theory as some kind of foundation of sand that has no merit in their "practical" world. Still others elevate theory, especially scientific theory, to the exalted status of unassailable unchangeable law of the universe... especially those theories which they personally favor. Now that we are investigating what science is, and how it is done, let's take a brief look at this thing called theory.

According to several dictionaries, a theory is:

1. The analysis of a set of facts in their relationship to one another;

2. A belief, policy, or procedure proposed or followed as the basis for action. "On the theory that...";

3. An ideal or hypothetical set of facts, principles, or circumstances—often used in the phrase "in theory";

4. The general or abstract principles of a body of fact, a science or an art (music theory);

5. A plausible or scientifically acceptable general principle or body of principles offered to explain phenomena (e.g., the wave theory of light);

6. A hypothesis assumed for the sake of argument or investigation;

7. An unproven conjecture or assumption;

8. A body of theorems presenting a concise systematic view of a subject; and an

9. Abstract thought or speculation.

Any statement of a theory must include as complete a statement as possible for the underlying assumptions (which could prove critical later on), the consequences, and predictions of observations that could be logically assumed to follow from the theory.

A theory is not a scientific discovery (only facts can be discovered), but rather is an intellectual construct derived from human ingenuity in the mind of the scientist. It is used to explain discoveries, but is not in and of itself a scientific discovery. Theories can be, and often are, proven wrong either on initial examination or later (sometimes centuries later).

The relationship between observations, theories, and laws are seen in our development of knowledge of the solar system of which Earth is a part. Prior to the 16th century, most Europeans believed that the Earth was at the center of the universe, a view put forth in ancient times by the Greek ruler of Egypt, Ptolemy. Nicolaus Copernicus, a Polish monk (ca. 1543), published the results of more than three decades of research that showed that the elliptical motions of the Ptolemaic system were best described by placing the Sun at the center of our solar system (the *heliocentric* view). Because of opposition from the Church, Copernicus withheld his speculations until he was on his deathbed. . . a precaution that his disciple Galileo failed to take to his ultimate regret.

The Danish astronomer Tycho Brahe (1546–1609) spent a lifetime making observations of the positions of the stars and planets. Brahe's work differed from earlier work because of the accuracy and precision of his measurements. Tycho Brahe, however, did not work up a viable theory or model concerning the data that he spent a lifetime gathering. That job was left to his disciple, Johannes Kepler (1571–1630), a mathematical physicist. Kepler discovered that the theories of Copernicus were not exactly correct because they disagreed with the data. By replacing Copernicus' circular orbits with ellipses, however, Kepler's model was able to explain the data and predict future positions of the planets. Kepler's statements of how the data behaved are collectively known, appropriately enough, as *Kepler's laws.*

Isaac Newton stood on the shoulders of Kepler, Brahe, Galileo, and Copernicus—he called it standing "on the shoulders of giants"—and derived his own laws of motion which are still celebrated in elementary physics classes today (even though later in the same course the relativity and quantum exceptions are discussed).

Figure 3-1 shows one model of how the theory and law-making process works. You may start out by making some observations, either in conjunction with a formal experimental research protocol, or by serendipity. These observations are then analyzed, and a model of how they work (or what they mean) is created. A tentative theory (as described above) is then developed from the model. From the theory predictions and hypotheses are derived, and these form the basis for an ordered observation, or experiment that makes additional observations. The process continues around and around the loop until the theories and laws derived are refined. But even then, the loop doesn't really stop; it only drops into very low gear, for there is always the possibility of a novel new observation coming along and restarting the process.

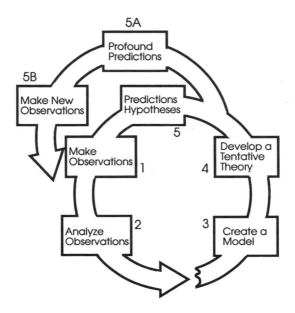

Figure 3.1.
The scientific method expressed as a never-ending cycle.

Sometimes, a tentative theory branches off into a different course because some of the predictions are truly profound. Einstein did one of these tricks using Max Planck's theory of energy quanta (1900), and won the Nobel Prize in Physics for his efforts.

Keeping Records

Beginning scientists are sometimes tempted to fall into the "doing" trap—i.e., *doing* their science without due regard for properly recording the experiments or observations. After all, science is a verb, isn't it? Doing experiments and making observations is fun, and it is what most readers are probably most adept at. But experiments without proper records are next to useless once their temporary entertainment value is exhausted. There are some very good reasons why you should be keeping proper scientific records of your work.

First, good records will allow you to recognize patterns that develop in your experiments or observations. Patterns can be an indication of underlying order, and one of the scientist's goals is to uncover hidden order in nature. It is almost axiomatic in science that such order exists, even in apparently disorderly processes. For instance, we find that the helter-skelter motion of heated gas molecules becomes orderly when understood statistically, or when large numbers of molecules are considered as a whole (e.g., gas pressure–temperature–volume relationships). Even modern chaos theory, which addresses what were once considered random processes, finds strange underlying order emerging out of chaotic situations.

One of the ways that scientific knowledge advances is through recognizing the patterns in nature. Those patterns may be subtle and not easily recognized when only hazy memories are available. Alternatively, nonexistent patterns sometimes seem to emerge from "remembered" (if murkily) data. If you keep good records, however, it is possible to compare later observations with earlier observations to see if there is anything in common that might indicate an underlying pattern. And those records are important whether you are an experimenter in a workshop or laboratory, or an observer in the field. The field observer might notice some unusual behavior of a species under certain circumstances. Later, if other individuals of that species are noted doing similar things, then it is possible to draw inferences about that animal's behavior patterns.

But you also must guard against being—dare we say it— "too scientific." How could you be in such a state? An example was seen in the case of the "Sliding-Board Crows." It seems that a rookery of crows had taken up residence in the vicinity of a university chapel. They took to congregating daily on the top ledge of the steeple tower, and then one by one they would jump off their perch to slide on their backs down the sloping surface of the steeple. When they came to the edge they would vigorously flap their wings to become airborne again. . . only to rise up to the ledge and try again when their next turn came around. After much observation, theory proposing, and scratching of heads, the biologists of the university finally answered the question of why the crows behaved in such a novel and bizarre manner. The apparently correct answer had nothing to do with survival of the fittest, or anything else. . . the Sliding-Board Crows engaged in their behavior *because it was fun*!

Experimental and observational science often rises and falls on statistics. It is from "stats" that you will be able to infer significance from results; statistics will permit you to better understand the phenomenon that you are studying. The usefulness of statistical calculations is only as good as the data used to make the calculations: remember, garbage in, garbage out. A good record of the conditions, variables, assumptions, and results of an experiment or observation will permit you to make such calculations later.

Another reason for keeping good records is that you will avoid doing needless repetitions of your experiments. Time is important, and some experiments are expensive. If you keep good records of your experiments, then you won't mistakenly do the same (or a too similar) experiment later on. Memories fade after a few months, and details become blurred. Your current bright idea may well be merely a residual memory of a failed experiment done ten years ago... which could be checked if records existed.

Repeating experiments is time-consuming and expensive. There are, however, situations where you *do* want to repeat an experiment. An unusual or unexpected result could be an "outlier" or "flyer," i.e., an experimental anomaly brought on either by random variation or by a fluke mistake in technique, observations, or materials. If you want to see if the result was real, then you will want to replicate the results by doing the experiment over again, perhaps with slight variations or perhaps identically with the original. Your records will come in handy in both cases because they will permit you to exactly replicate the original experiment. Alternatively, they give you a basis for fine-tuning the experiment to see if the results are improved.

> **OUTLIER'S—BEWARE**
>
> "Outliers" ("flyers") are data points that are far removed from the main grouping of points. For example, an outlier may be far from the average. Many experimenters are all too eager to simply drop these data, and are often encouraged to do so. But apply a little critical thinking first! The outlier may indicate an unknown phenomenon, or some previously unsuspected source of correctable error in the experiment. Don't just glibly toss out data.

Perhaps more important than avoiding repeating the same or similar experiments is the ability to extend the work. Your knowledge of science is extended by building on past work. As mentioned, the great Isaac Newton stood on the shoulders of those scientists who worked on related (or the same) problems before him. A scientist can also stand on his or her own shoulders by carefully redesigning and extending the scope of past experiments to further develop the topic. Studying the records of past experiments will suggest topics for further study of the same general area.

When planning further experiments, or when contemplating another field trip, you can figure out what was done before, what was overlooked before, and what is needed to correct those oversights.

You will get a lot more out of your science activities if you can go about them in an orderly, planned manner—and the records will be your planning road map. But there can also be other good reasons for keeping detailed records.

First, you may discover something that is scientifically important and want credit for the discovery. If that discovery is adequately documented, then the scientific community will grant you recognition for it. Even amateur scientists can receive credit for their discoveries. Who knows, science student Linda Zliltz may go down in history for discovering the Zliltz Effect in her basement laboratory. Doesn't happen, you say? Look at how many comets are named after amateur astronomers! If you are interested in field biology, then you may well be the first person in your region to note that the beautiful *Flapitzwingus Whatzitcallitz* has apparently migrated from its haunts in the south into nearby forests. It happens! While your fame might not be national in scope, or win you a Nobel Prize, it can nonetheless be very satisfying to be recognized for a small contribution—even if only of purely local interest.

Second, in trying to solve a problem in experimental science you might design (invent!) an instrument to make a measurement, or a device that performs a process or does some other chore. Checking with a patent attorney, you find that your invention has more than a little merit, and that it is patentable. That gives you 17 years of exclusive

---

THE RANGE OF A SPECIES

Wisdom. The range of any species is directly proportional to the range of science students searching for it.

commercial exploitation of your work. In order to realize the benefits, however, you will have to prove that the invention is yours and that it works. The records that you kept in your research work will be valuable towards establishing your claim to a patent in the first place, and in defending the patent later on if it is either challenged or infringed.

What Kind of Data to Record?

There is a running debate (argument?) on how much data, and what kind, should be recorded in a scientist's notebook. Some believe that only the most pertinent data should be recorded, the rest being irrelevant. Others claim that there is no such thing as *too much* data. The only real limit is your own patience, and how much effort you want to expend in keeping records.

For any given situation, however, there is probably some optimum set of data that should be collected and recorded. When deciding on what to record, keep in mind that data is the basis for information (indeed, information has been called "data with order and context imposed"), and that the key to information is the *inform* part. In other words, the data collected ought to be capable of informing you of something or other. It is the nature of the experiment or observation that will determine what, and how much, to record. We can, however, offer some typical examples. (Samples of pages from a scientist's notebook—courtesy of Forrest Mims—are provided on pages 44 and 45.)

A field biology or geology enthusiast will want to take a notebook on trips to record what they find. Depending on the situation, record the time of day, temperature, the weather conditions, and other physical facts that are found with the actual observation. For example, when observing some particular species of animal, its behavior might be related to some environmental condition in the immediate area.

You will want to record the location where the observation was made as precisely as possible. I've known rockhounds, field geologists, and amateur archaeology and history buffs who lament not properly recording the location of some find. . . and they were unable to locate it again! For example, one friend of mine located an outcropping of a semiprecious stone. While the intrinsic worth was only moderate (it probably wasn't worth commercially mining), when it was finally identified it was believed to be the first time this mineral was found in Virginia. Unfortunately, when trying to lead other members of his nature club back to the spot where the sample was found, he became confused and never did relocate the source! *Pity.* A few accurate, and easily followed, directions would have been valid data in his geology field notebook.

It is well recognized that the results of laboratory experiments should be recorded. What is often overlooked, however, is that experiments often have inputs, initial conditions, assumptions, or other details besides the results. These should be recorded, where appropriate.

Record the configuration of any experimental apparatus used. In other words, provide a schematic or drawing of how

LAND POSITION FINDING

The problem of relocating a discovery dogs many field scientists, especially in terrain that looks alike for many miles around. One solution that is open to researchers in much of the United States and Canada is to use the mariner's Loran-C navigation radio system. Hand held Loran-C receivers are available that allow you to pinpoint your latitude and longitude to a precision of 50 feet or so. They will also allow you to generate orienteering waypoints for trekking from some present location to the desired location. See any marine electronics dealer for details on handheld Loran-C receivers. Readers in other countries also have Loran-C (or alternatives) available, but must buy receivers from local sources because of their uniqueness.

you "lashed up" the apparatus used to do the experiment. Also describe the experiment in prose, and provide yourself with a step-by-step procedure, if that is appropriate. This will allow you or others to replicate the experiment later on. If you use instruments or other equipment, then record the model number of each item. And if multiple versions of the same instrument are available, then record the serial number or other identification of the one used. Sometimes, anomalous results in an experiment are caused by a subtly malfunctioning instrument, rather than being a new find in nature. Remember Milligan's Law, cited in Bob Pease's book *Troubleshooting Analog Circuits*: "When you are taking data, if you see something funny, Record Amount of Funny."

You might also be well advised to record ambient room temperature, humidity, or other factors that might influence the results of your experiment. If an electrical apparatus is used, then the power line voltage or the settings of controls on the front panel might prove interesting later. If you use chemicals or other disposable supplies, then it might also be important to record the name of the manufacturer, date of manufacture or expiration dates (as appropriate), lot number, manufacturing code or other pertinent details. These may seem like trivia, but if the result is unexpected, then the solution to "why" may well be found in some trivial but overlooked detail that you recorded in your notebook.

When recording numerical data, be sure to record a sufficient number of observations to draw conclusions. A single-point observation is of little or no use. While it may be interesting, it has no significance unless a lot of other observations are recorded. For any given situation, there will be a statistically significant number of results that have to be obtained. In these cases, a large number (20 to 30 minimum) of repetitions of the experiment need to be done. Such data is best recorded in tabular form so that the patterns can be recognized, or statistical calculations performed.

The types of data discussed above can be awfully sterile, and while they are interesting and necessary parts of science, there is also room for a "diary" of personal impressions and

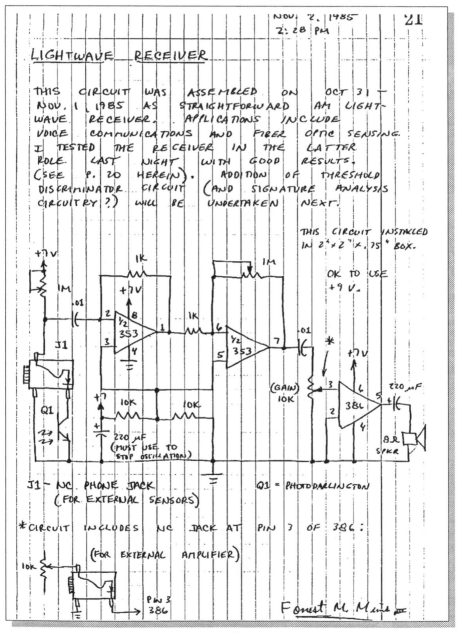

Sample pages from a scientist's notebook. Date and time are recorded on each page. It's also a good idea to initial or sign each page as it is used.

observations. Some of these might prove to be scientifically useful in some situations, but they almost invariably add a personal touch to your work. Keep in mind that impressions

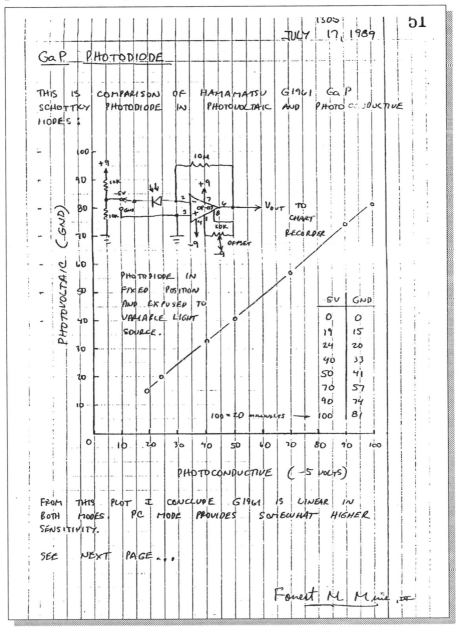

1505

JULY 17, 1989   51

GaP PHOTODIODE

THIS IS COMPARISON OF HAMAMATSU G1961 GaP SCHOTTKY PHOTODIODE IN PHOTOVOLTAIC AND PHOTOCONDUCTIVE MODES :

PHOTODIODE IN FIXED POSITION AND EXPOSED TO VARIABLE LIGHT SOURCE.

V_{out} TO CHART RECORDER

100 = 20 millivolts →

PHOTOCONDUCTIVE (-5 VOLTS)

5V	GND
0	0
19	15
24	20
40	33
50	41
70	57
90	74
100	81

FROM THIS PLOT I CONCLUDE G1961 IS LINEAR IN BOTH MODES. PC MODE PROVIDES SOMEWHAT HIGHER SENSITIVITY.

SEE NEXT PAGE ...

Forrest M Mims III

are not always good science (being so-called "anecdotal data"), but they are the poetry that makes science fun. Besides, lurking in the prose impressions, alongside the cold, hard "real" data, might be some element of your impressions that contains the seed that sparks further research ideas at a later time. In addition, many years later, when you are approaching your dotage and are no longer able to hike over hill, dale, and rocky mountainside, those impressions and personal comments may be immensely satisfying to review while rocking in your chair on the front porch.

What Kind of Record Books?

There is no "required" format for your records, although there are some guidelines. After all, the records are kept to serve *you*, and therefore should be in a format that you can most easily use. The record can be a formal scientific notebook, or a simple informal diary that has meaning only to you.

The form of your scientific activity is also important when deciding the type of notebook or diary to use. A large notebook is well suited to indoor laboratory activities where there is either plenty of desk space or a handy dining room table that can be used for the paper-work part of science. But if you are doing field work, then a large format notebook could be a real hassle. A smaller format might be easier to live with out in the boonies.

The accepted format for formal scientific records is the bound (not loose-leaf) notebook with numbered pages. The numbering should be printed onto each page by the manufacturer, not by the owner. Furthermore, the best notebooks are the type in which the numbered pages cannot be either removed or torn out without leaving a tell-tale trace. The more acceptable notebooks use stitch binding methods rather than glue binding, although the right kind of glue binding is also usable if pages cannot easily be removed. The goal is to have a rugged, long-lasting notebook. The bound-and-numbered format makes the notebook a good sequential record of your experimental progress. Keep in mind, however, that these notebooks are expensive, and your needs might not require such formality. You might, for example, want to save money by numbering your own pages, even though it is not formally correct.

Several different forms of bound notebooks are available for scientists. They can be purchased at stationary stores (some types at least), school supplies outlets, engineering, drafting or scientific supplies stores (those that cater to professionals), and from local and college university book stores.

Composition Books. These notebooks are found in school supplies sections of drug stores, stationary stores, and paper goods stores. They are the kind that have a black-and-white speckled hard cardboard cover, and ruled pages to guide handwriting. Some of them are found with printed numbers on each page, and are the preferred types. These are the notebooks that high school and freshman college science instructors often require students to use for the laboratory sessions of their courses.

Record Books. Nearly all stationary stores sell bound books for business records. Columnar Books are used for accounting, Payroll Books are for keeping up with the money paid to workers, and Records Books are used for general records keeping. It is the latter that interest us. The Records Book is a bound book, with numbered and ruled pages similar to the composition notebook discussed above. Some of these books are the same approximate size as Composition Books, while others are a bit smaller. Most of them, however, are larger. I've seen Records Books that are 24 inches high by 16 inches wide (with 350 pages), but they are an exception on the other extreme. Others are only 6 by 9 inches, and are easily tucked into a backpacker's knapsack. Popular commercial Records Books are the *National Account Books* Cat. No. 57-211 (150 pages) and Cat. No. 57-231 (300 pages). These bound books have 150 and 300 numbered pages, respectively, each page being about 7.8 by 10.25 inches, and shaded light green for "eye ease." Don't be shocked when you find out how much these books cost, by the way. But consider how much use the book will be over a long period of time.

Engineering Laboratory Notebooks. These are usually paper bound notebooks of large size, with quadrilled graph paper ruling. Engineers use them in their design work to record progress, results, and false starts. They are very similar to. . . .

Scientist's Laboratory Notebooks. These notebooks are bound, have numbered pages that are made of either quadrilled or ruled paper. Many of these notebooks, as well as some engineering notebooks, come with two sets of pages (typically yellow and white). One set is permanently bound into the notebook, and the other, which carry a duplicate set of page numbers, are perforated so that they may be detached and stored elsewhere. A piece of old-fashioned carbon paper may be attached to the back of the book. It is placed between white and yellow pages of the same number so that a copy is made of all writing on the page. The perforated copy may be detached and stored in another location to guard against loss. There is more than one scientist who lost a valuable notebook in an airport or taxi-cab. . . but fortunately had another copy back home.

Blank Books. Some sketch books, diaries, or other forms of bound book are available with bound pages. If these do not have numbered pages, then they may be less desirable for those who require or want formal records. On the other hand, if they do contain the required numbered pages, or you are happy with handnumbering or unnumbered pages, then these books are a cheap way to get started. They are available in sizes from hip pocket minibooks to "two-person carry" megabooks. Some of them have heavy paper covers, while others look like hardbound books. Look for these books in artists' supplies stores.

Keeping the Notebook

Each entry must be properly dated, and if appropriate, the time of day recorded. If these data were not included, an ex-lab instructor of mine would reject the student's work by writing a big red crayon slash mark across each of the offending pages. *The vandal!*

When you write dates use the "military" method. In other words, a date is written *day-month-year*, with the month spelled out or abbreviated. Using the spelled out version prevents ambiguities from arising. For example, "6/9/91" could mean "June 9, 1991" or "September 6, 1991." If you make it your habit to use the correct format ("9 June 1991" or "6 Sept 1991"), then there will be no question as to what day was meant.

Similarly, times can be ambiguous so use the 24-hour clock system (also preferred by the military). After all, "8:30" could mean either a bit after breakfast (8:30 AM) or a bit after dinner (8:30 PM), but "0830" and "2030"—the 24-hour clock equivalents—are unambiguous. It is also sometimes useful to indicate the time zone in which the time was recorded.

If you are attempting to make simultaneous measurements with other people in different parts of the country (or the world, for that matter), then you might want to keep the time in *Greenwich Mean Time* (GMT), also called *Universal Time* (UT)—i.e., the time in Greenwich, England, at the prime longitudinal meridian. A widely used shorthand for UT/GMT time readings uses the letter Z as a suffix. For example, "1800Z" means 1800 hours (i.e., 6 PM), Greenwich Mean Time. The "Z" suffix is sometimes called "zulu time," and is derived from the lingo of radio operators. Pronouncing "eighteen hundred hours zulu time" is a little easier than the alternatives, and for Morse code operators the "Z" suffix is a lot less "brasspounding" than spelling out *Greenwich mean time*.

It is also a good idea to initial and date each page of the book as you use it, especially if more than one person has access to the notebook. Some laboratory notebooks come with a date/initial block printed on each page. Subsequent entries, error corrections, or new information recorded on any given page should be separately dated and initialed so that it is clear to subsequent readers when a datum was entered.

One of my old prof's pet peeves was pencil entries. If you are attempting to keep a formal record, then all entries in your notebook should be done in indelible ink—never in pencil or any other erasable media. If you make a mistake, then cross through it and record the correct data. Otherwise, erasures may corrupt your record in the minds of even the most fair and objective critics. Included in this ink record are notes, calculations, observations, questions that arise, and any other detail that comes up.

Another advantage of indelible media is that "errors" and "mistakes" often turn out to be pertinent later on, as they may indicate a problem. Bob Pease, writing in his column "Pease Porridge" (*Electronic Design* magazine), tells the story of a technician at his company's Texas plant. She noted a log book reading of phosphine measurements, but knew that phosphine was not used in the process being run at the time. On initial investigation, the engineers dismissed her concern as a recording error. Phosphine wasn't required for the process, so it obviously couldn't be running. But the technician wouldn't let it go, and on her own shift noted that phosphine was flowing. . . and demanded the engineers do their job. It seems that a software glitch. . . [old story!]. If the original "funny" entry is merely erased, then an important opportunity to correct a real problem is lost. Maybe there is a corollary to Milligan's Law: Record Amount of Funny, and *then kick-butt to find out "Why Funny?"*

Graphs, photographs, paper strip charts, computer printouts or other output media from instruments are handled a little differently. In each of these cases, tape or permanently paste (never staple) the record onto a page in the notebook, and provide handwritten (ink) annotation as to what it is. Some people initial and date these secondary records before placing them in the notebook.

Also considered a good idea is to initial and date photographs. Try to find an area on the margin where the initials can overlap an unimportant portion of the photo emulsion so that it will be obvious that it is not a replaced or doctored photograph (this is the accepted method of signing photos for passports in some countries, by the way).

Finally, in science the main difference between fairly using a good literature search and evil, dirty, filthy plagiarism is attribution! The difference between a scholarly researcher and an intellectual thug is a footnote. Even the best scientists, like Newton, stand on the shoulders of giants. In many situations, you are likely to adopt or adapt ideas from others. These borrowings can be noted in your laboratory notebook. When you write the idea into a science report, a term paper, an article, or other scientific communication, then you can avoid a lot of problems—and heated accusations—by providing a properly written source note either in the text or as an endnote (or footnote).

Conclusion

One of the formal rules of evidence used in courts, in scientific circles, and by logicians is that a missing record indicates an event did not take place, *if* it is normally customary to record such events—your subsequent protestations to the contrary notwithstanding. So if you have a certifiably bright idea some day, it could be taken away from you by your sloppy records keeping. *Keep up a good experimental lab notebook!* Besides, it's fun and adds to your enjoyment of your science activities over the years.

5

Some Mathematical Basics

OING SCIENCE is certainly stimulating and a lot of fun, but you must have certain basic arithmetic skills to do science right. Many readers may already know the material in this chapter, and would find it trivial. Others, however, are not so familiar with basic subjects such as significant figures, scientific notation, and solving problems by dimensional analysis. Experts are encouraged to review the material, but if that is a waste of time in your case please feel quite free to "fast forward" to another chapter.

Significant Figures

Much of our common experience deals with exact numbers of things: 5 stamps, 127.50 dollars, and 4 people. These items can be counted and an exact numerical representation provided; all figures are significant in such cases. But in other situations, you may take measurements that are subject to errors. For example, you might measure the height of a person as 67 inches, 68 inches, or 69 inches depending on how straight the person stands. Or what about the answer you give when asked a person's weight? The scale may register 162 lbs, but one of the balance weights may not be perfect, or perhaps the scale dial at rest sticks a little off zero. How many people do you know who own a perfectly accurate watch that never needs to be reset? None! These flaws are implicitly resolved when we apply the concept of *significant figures* to the measurements. This concept demands that we impute no more precision to a measurement or calculation than the natural reality of the situation permits.

The counting numbers (1, 2, 3, 4, 5, 6, 7, 8, and 9) are always significant. Zero (0) is significant only if it is used to indicate exactly zero, or a truly null case. It is not significant if it is used merely as a place holder to make the numbers look nicer on the printed page. For example, if the number is properly written, then "0.60" means *exactly* 6/10, and is not approximately 0.6; the second zero used here is significant. If the number is written "0.6" then we may assume that it means 6/10 plus or minus some amount of error or uncertainty.

When we use numbers to indicate a quantity, then the concept of significant figures becomes important. For example, "16 gallons" has two significant figures, but can be taken to mean that the quantity of liquid is somewhere between 15 and 17 gallons. But if our measurer of liquids is better, then we might write "16.0" gallons to indicate precisely 16 gallons plus or minus a smaller error; for instance, perhaps the real measure is between 15.9 and 16.1 gallons. Consider a pressure gauge that is guaranteed to an accuracy of ±5%. A reading of "100 Torr" has three figures, meaning that the actual pressure is between [100 − 5%] and [100 + 5%], or 95 to 105 Torr (two significant figures).

Consider a practical situation. An experiment uses a digital voltmeter to measure an electrical potential difference of exactly 15 volts. The instrument reads from 00.00 to 19.99 volts, with an accuracy of ±1%. In addition, all digital meters have a ±1-digit error in the least significant position; this problem is called *last-digit bobble*. For the digital voltmeter in question:

19.99

Most significant digit _____↑   ↑_____ least significant digit

The last-digit bobble problem means that a reading of 15.00 volts could be any value between "15.00 – 00.01" (14.99) volts, and "15.00 + 00.01" (15.01) volts. In addition, the error of 1% means that the actual voltage could be ±(15 x 0.01) = ±0.15 volts. Thus, the actual voltage could be (15.00 – 0.15) volts to (15.00 + 0.15) volts, or a range of +14.84 to +15.16 volts.

If both errors are minus:

Reading: 15.00 volts
            –0.01 volts
            –0.15 volts
            _____

            14.84 volts

or, if both errors are positive:

Reading: 15.00 volts
            +0.15 volts
            +0.01 volts
            _____

            15.16 volts

Significant figure errors can be propagated in a calculation. A rule to remember is that the number of significant figures is not improved by combining the numbers with other numbers. For example, multiplying a significant digit by a nonsignificant digit yields a result that has at least one nonsignificant digit. It is often the case that the number of significant figures decreases. Suppose we measure a voltage (V) as 15.65 volts, and a current (I) in the same circuit as 0.025 amperes. The power (P) is the product VI. Let's take that product, placing a little hat (^) over each digit that is not significant, and then carry that notation down wherever a nonsignificant digit is a factor with another digit.

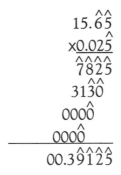

As can be seen, only the "3" and one of the leading zeros are significant. Thus, we are claiming more precision than is truly available if we list the power as "0.39 watts," when the "9" is not significant. We might be better advised to list this value as "0.4 watts."

The reason that scientists and engineers make such a fuss over significant figures is that it is both bad form, and potentially dangerous under the right circumstances, to claim more precision than is truly the case. For this reason, we typically limit the figures to the number of decimal places for which a reasonable expectation of physical reality exists.

Significant figures rules were a little easier to understand and use in the days when scientists and engineers made calculations on a slide rule. But in this age of $12 scientific pocket calculators and personal computers everywhere, the distinction often gets lost. Consider a simple electrical problem. One version of Ohm's Law states that the current flowing in a circuit is the quotient of the voltage (V) and the resistance (R). Suppose 10 volts is applied to a 3-Ω resistance. According to my $12 pocket scientific calculator, the current is 10/3 = 3.333333333 amperes. Does anyone really think that their ordinary, run-of-the-mill, laboratory ammeter can measure to within 10^{-9} amperes? In most cases, we would be bragging out of school to claim more than 3.33 or 3.333 amperes using very good meters with recent calibration stickers on them!

Scientific Notation

Scientific notation is a simple arithmetic shorthand that allows you to deal with very large numbers using only a few digits between 1 and 10, and powers-of-ten exponents. The form of a number in scientific notation is:

$$\underset{\text{Numbers}}{\underbrace{\text{n.ij}}} \times 10^{\overset{\text{Exponent}}{x}} \quad \text{Base 10}$$

For example, if the age of a college physics professor is 47, it could be written:

$$\text{Prof's Age} = 4.7 \times 10^1 \text{ years}$$

(Note the units "years" in the preceding equation. The specification of a value is never complete if the units are not included; "47" or "4.7×10^1" are *not* the same as "47 *years*" or "4.7×10^1 *years*." The only exception is when the quantity is non-dimensional).

When the exponent is negative, it is the same as saying $1/10^x$. In other words:

$$10^{-x} = \frac{1}{10^x}$$

Some of the standard values in power-of-ten notation, along with their respective prefixes for use with units, are:

1/1,000,000,000	=	$0.000000001 = 10^{-9}$	(pico)
1/1,000,000	=	$0.000001 = 10^{-6}$	(micro)
1/100,000	=	$0.00001 = 10^{-5}$	
1/10,000	=	$0.0001 = 10^{-4}$	
1/1,000	=	$0.001 = 10^{-3}$	(milli)
1/100	=	$0.01 = 10^{-2}$	(centi)
1/10	=	$0.10 = 10^{-1}$	(deci)
1.0	=	10^0	
10	=	10^1	(deka)
100	=	10^2	
1,000	=	10^3	(kilo)
10,000	=	10^4	
100,000	=	10^5	
1,000,000	=	10^6	(mega)
1,000,000,000	=	10^9	(giga)

(Note that 1,000,000,000 is called "one *billion*" in the U.S., but "1000 million" in England and most of the rest of the world. The term *milliard* was once applied to 10^9. To be one billion outside of the U.S. the number would have to be 1,000,000,000,000 (10^{12}). Perhaps we should label 10^{12} *billiards* so that Carl Sagan could go on TV and talk about "billiards and billiards of worlds.")

Scientific notation is especially appealing when dealing with numbers for which there are reasonably only a few significant figures. If we measure a brain wave surface potential as 143.6 µV (microvolts), then we may prefer to represent it as 1.44×10^{-4} volts.

The prefixes above are used to subdivide units. For example, *milli* means 1/1000 (or 0.001), so a *milli*meter is 0.001 meters. Similarly, *kilo* means 1,000, so a *kilo*meter is 1,000 meters.

Rules for Scientific Notation Arithmetic

I.   Addition and Subtraction: Force the exponents to be equal, and then add the numbers.

II. Multiplication: Multiply the numbers but *add* the exponents.

III. Division: Divide the numbers, but subtract the exponents.

Of course, if you have a pocket scientific calculator, then the EXP button takes care of these little chores for you.

Solving Problems the Dimensional Analysis Way

Science problems are often little more than applications of a few basic principles, but frequently loom much larger to the student. A disgustingly clever technique is to use what is called *dimensional analysis, units conversion,* or *factor line* analysis.

I have seen student nurses fret needlessly over drug dosage arithmetic exams, a necessary major waypoint in their academic careers (usually administered early in the freshman year). The root problem is not really the lack of mathematics ability, for the only skills needed are the four basics: addition, subtraction, multiplication, and division. What they needed was a coherent problem-solving system. Dimensional analysis provides that framework.

In order to use this method, you need to establish a ground rule that is never to be disregarded: *Units are a part of the problem, and are treated mathematically the same as numbers.* In other words, when a unit or dimension is appropriate, always use it. When a length is used, then the unit is as much a part of the length description as the number. A length of "one foot" is properly "1 ft," not simply "1."

Another thing that must be remembered is the simple (and almost self-evident) fact that things equal to the same thing are equal to each other. Consider an example. A table of conversion factors will tell us that a mass of one kilogram (1 kg) has a weight of 2.2 lbs (we can use a unit of mass as a unit of weight when we realize that the "mass weight" is nothing more than the attraction of the Earth on the mass by the force of gravity). Therefore:

$$1 \text{ kg} = 2.2 \text{ lbs}$$

The basic underlying physical fact that validates this equation is that a platform scale will balance exactly if a 2.2-lb weight (or other type of 2.2-lb force) is on one dish and a 1-kg weight is on the other dish. Both expressions, "1 kg" and "2.2 lbs," represent identical amounts of material. But what would happen if we were to divide both sides of the equation by 2.2 lbs? This is legal arithmetic as long as we do the same thing to both sides of the equation.

$$\frac{1 \text{ kg}}{2.2 \text{ lbs}} = \frac{\overset{1}{\cancel{2.2 \text{ lbs}}}}{\underset{1}{\cancel{2.2 \text{ lbs}}}}$$

Notice that the "2.2 lbs" is canceled out (actually it divides out, but that process is commonly—if erroneously—called "canceling"), so the statement reduces to:

$$\frac{1 \text{ kg}}{2.2 \text{ lbs}} = \frac{1}{1} = 1$$

This step is not a mathematical flim-flam (take note, please, anyone who knows that $1/2.2 = 0.455$ and $0.455 \neq 1$), especially when the "=" symbol is taken to mean "the same as..."

Similarly, if we had divided both sides of the original equation by 1 kg instead of 2.2 lbs, then the result would be:

$$\frac{\dfrac{1}{\cancel{1 \text{ kg}}}}{\dfrac{\cancel{1 \text{ kg}}}{1}} = \frac{2.2 \text{ lbs}}{1 \text{ kg}}$$

or,

$$\frac{2.2 \text{ lbs}}{1 \text{ kg}} = \frac{1}{1} = 1$$

Now we have two different expressions involving 1 kg and 2.2 lbs, and both of them are equal to 1: *Things equal to the same thing are equal to each other*:

$$\frac{2.2 \text{ lbs}}{1 \text{ kg}} = \frac{1 \text{ kg}}{2.2 \text{ lbs}} = 1$$

Mathematically, it is permissible to substitute one of these expressions for the other without changing the problem and making a mistake.

We also know that we can always multiply a quantity by 1 without changing its value:

$$2 \times 1 = 2$$

$$9 \times 1 = 9$$

$$A \times 1 = A$$

$$4 \times \frac{2}{2} = \frac{8}{2} = 4$$

The above relationships are easy to see. What is a little harder to see, but is nonetheless true, is the following:

$$4 \text{ kg} \times \frac{2.2 \text{ lbs}}{1 \text{ kg}} = 8.8 \text{ lbs} = 4 \text{ kg}$$

The principle is the same: 8.8 lbs is still the same quantity as 4 kg.

While these relationships seem like a trivial matter, they represent one of the most powerful tools in the tool kit for solving problems. Let's solve a sample problem to see how it works. Assume a registered nurse (RN) needs to convert a mass of 3 grains (used with older pharmaceuticals) to the equivalent number of milligrams. We know two facts:

(1) There are three grains (3 gr)
(2) 60 mg = 1 gr (from a conversion table)

The fundamental trick in dimensional analysis is to multiply given amounts, such as (1) above, by conversion factors equal to one until only the desired units are left uncancelled in the answer. In this particular problem, we have a conversion factor equal to one involving both milligrams (the desired units) and the given units (grains):

$$\frac{60 \text{ mg}}{1 \text{ gr}} = \frac{1 \text{ gr}}{60 \text{ mg}} = 1$$

We select the one which will cancel the grains unit in the numerator. A unit is canceled by the same unit placed so that the two will divide out ("cancel"). That is to say, a grain in the numerator can only be canceled by another grain in the denominator. Therefore...

$$3 \text{ gr} \times \frac{60 \text{ mg}}{1 \text{ gr}}$$

The grains cancel out, leaving us with:

$$3 \times 60 \text{ mg} = 180 \text{ mg}$$

We know that the answer is correct by using a four-point check method:

(1) Is the conversion factor correct? (Does 60 mg really equal 1 gr.?)

(2) Is the arithmetic correct? (Does 3 × 60 really equal 180?)

(3) Are the numbers reasonable? For example, we know that grains is a larger unit than milligrams, so the answer ought to contain a number larger than the conversion factor. If the answer had been 20 mg, then we know it was set up wrong because 20 mg is less than 60 mg.

(4) Are the desired units (e.g., milligrams in the above example) the only ones left uncancelled in the answer?

Let's look at another example. A solid substance is dissolved in a liquid base such that 1 mg is dissolved in 2 cc, and is labelled "1 mg = 2 cc." How many cc are required in order to obtain 5 mg of the material? Questions like this are routinely found in chemistry, medicine, nursing, and the life sciences.

$$5 \text{ mg} \times \frac{2 \text{ cc}}{1 \text{ mg}} = \frac{5 \times 2}{1} \text{ cc} = 10 \text{ cc}$$

The same type of problem is often seen with the dissolved material listed as "1 mg per 2 cc." The "per" word is nothing more than a verbal division symbol: "1 mg per 2 cc" means the same thing as "1 mg = 2 cc," or "1 mg in 2 cc," or "1 mg/2 cc."

Keeping with our medical examples, let's look at an example involving the administration of an intravenous solution. A patient is to receive 3,000 cc of an IV solution over 24 hr. The mechanism used to administer the solution has a drop factor of 15 drops (gtt) per cc of solution. We need to find the drip rate in drops per minute (gtt/min):

$$\frac{3,000 \cancel{cc}}{24 \cancel{hr}} \quad \times \quad \frac{1 \cancel{hr}}{60 \text{ min}} \quad \times \quad \frac{15 \text{ gtt.}}{\cancel{cc}} \quad = \quad ???$$

$$\left(\frac{3,000 \times 15}{24 \times 60} \right) \frac{\text{gtt}}{\text{min}} \quad = \quad ???$$

Because only the desired units, gtt/min, are left uncancelled in the answer, we may now carry out the arithmetic to get the numerical answer:

$$\left(\frac{3,000 \times 15}{24 \times 60} \right) \frac{\text{gtt}}{\text{min}} = 31 \frac{\text{gtt}}{\text{min}}$$

Notice that we left the arithmetic unworked until the very last step. Always work with the units and forget the arithmetic until all factors have been included and *only the desired units are left uncancelled*. Then, and only then, turn the calculator crank and spit out the lean, freshly ground, Grade-A answer.

Turning to another field for our next example, let's look at the sensor sensitivity problem. A certain device called a Wheatstone bridge pressure transducer is excited by a 5.00 volt dc (direct current) potential (V), and produces an output voltage that is a function of the applied pressure, the excitation potential, and the sensitivity factor (φ). Let's assume that the sensitivity factor for a specific device is 5 millivolts (mV) per volt of excitation (V) per Torr (T) of pressure:

$$\varphi = \frac{5 \text{ mV}}{V \cdot T}$$

The output voltage (V_o) is found from:

$$V_o = \varphi V T$$

So, with V = 5 volts, what is the output voltage when a 400-Torr pressure is applied?

$$V_o = \left(\frac{5 \text{ mV}}{V \cdot T} \right) (5.00 \text{ V}) (400 \text{ T})$$

$$5 \text{ mV} \times 5 \times 400 = 10,000 \text{ mV}$$

And how many volts is 10,000 mV?

$$10{,}000 \text{ mV} \times \frac{1 \text{ V}}{1000 \text{ mV}} = 10 \text{ V}$$

New problem. The drug *Dopamine* is prescribed by physicians by the number of micrograms (μg) of drug per kilogram of body mass per minute of time. For example, a physician's order might be 5 μg per kg per min. The drug often comes from the pharmacy dissolved 400 mg in 500 cc of IV solution. Because the drug causes large changes in blood pressure for small changes in the drug administration rate, the nurse will typically use a microdrip chamber mechanism that offers 60 gtt/cc, rather than the normal 10 or 15 gtt/cc. What drip rate is needed to deliver the drug at the prescribed rate to a 90-kg (198-lb) patient?

$$\frac{5 \text{ μg}}{\text{kg} \cdot \text{min}} \times 90 \text{ kg} \times \frac{1 \text{ mg}}{1{,}000 \text{ μg}} \times \frac{500 \text{ cc}}{400 \text{ mg}} \times \frac{60 \text{ gtt}}{\text{cc}} = \text{???}$$

$$\frac{5 \times 90 \times 1 \times 500 \times 60 \text{ gtt}}{1000 \times 400 \text{ min}} = \text{???}$$

Only gtt is left uncancelled in the numerator, and minutes is left in the denominator. Our units, then, are gtt/min, which is what we want in the final answer. This tells us that it is time to turn the crank (or punch the calculator) to find the numerical part of the answer:

$$\frac{5 \times 90 \times 1 \times 500 \times 60 \text{ gtt}}{1000 \times 400} = 34 \; \frac{\text{gtt}}{\text{min}}$$

Now for an example I saw in college, during which time one of the professors, the author of a handbook of units conversions, was selling his book to students. The problem is this: A bicyclist pedals at a rate of 4,200 furlongs per fortnight (4,200 FL/FN) for a period of 20 minutes. How many yards did he travel? Hmmm, consulting the professor's book (a smart move) showed that 1 FL = 220 yards, and 1 FN = 14 days, so:

$$x = VT \text{ (distance = velocity} \times \text{time)}$$

$$x = \frac{4,200 \text{ FL}}{\text{FN}} \times (20 \text{ min})$$

$$= \frac{4,200 \text{ FL}}{\text{FN}} \times 20 \text{ min} \times \frac{1 \text{ FN}}{14 \text{ days}} \times \frac{1 \text{ day}}{24 \text{ hr}} \times \frac{1 \text{ hr}}{60 \text{ min.}} \times \frac{220 \text{ yd}}{1 \text{ FL}} = ???$$

Now that only yards are left uncancelled...

$$\frac{4,200 \times 20 \times 1 \times 1 \times 220 \text{ yd}}{14 \times 24 \times 60} = 917 \text{ yds}$$

Beware of professors who write odd-Schtick books!

The technique of dimensional analysis is a powerful method for solving problems. Just remember that the units are as much a part of the problem as the numerical part.

Σ Notation

Throughout much of this book you will see arithmetic equations using the upper-case Greek letter *sigma* (Σ). This symbol is used in math to denote "add 'em all up together." For example, the statement

$$A = \sum_{i=1}^{n} x_i$$

means "add together all of the values of x from the first to the nth. If there are five values of x (i.e., $n = 5$), then:

$$A = \sum_{i=1}^{5} x_i$$

This equation is the same as saying:

$$A = x_1 + x_2 + x_3 + x_4 + x_5$$

EXAMPLE

The following eight values ($n = 8$) of x are known: $x_1 = 2.5$, $x_2 = 6$, $x_3 = 4.2$, $x_4 = 3$, $x_5 = 1.75$, $x_6 = 4$, $x_7 = 2.6$, and $x_8 = 5$. Find "A" in the following equation:

$$A = \sum_{i=1}^{8} x_i$$

$$A = 2.5 + 6 + 4.2 + 3 + 1.75 + 4 + 2.6 + 5 = 29.05$$

In some problems, the sigma notation is used to find an average value (see Chapter 6). In those cases, you will see the expression written in the form:

$$\bar{x} = \frac{1}{n} \sum_{i=1}^{nx_i}$$

or,

$$\bar{x} = \frac{\sum_{i=1}^{n} x_i}{n}$$

The symbol x-bar ($\bar{x}$) denotes "average of x."

You will also see the lower-case *sigma* (σ) used in this book. It is not to be confused with Σ, because it means something different. You will find σ used to denote the standard deviation, covered later in another chapter.

Integration and Differentiation Operations

The mathematical processes of integration and differentiation are used extensively in scientific calculations. These processes are part of the mathematics known as the calculus. Although you will not need a calculus course to understand the material in this book, the concepts are used occasionally, so it's a good idea to become familiar with the basic ideas and notation.

Integration involves finding the area under a curve, and the result is called the *integral*. If the curve is a horizontal straight line, such as line AC shown in Fig. 5-1a as running from x_1 to x_2, then the task is easy. The area under the line is a rectangle or square, and its area is the product

$$A = (y_1 - 0)(x_2 - x_1)$$

(Note: if the base of the rectangle is at a location other than zero, then the "0" is replaced with the new value of y).

If the curve defines some regular geometric shape, we can either construct or look up the formula for the area of the figure and replace the parameters with the correct values of x and y. For example, for line AB (shown dotted in Figure 5-1a) it is possible to break the area into a rectangle (x_1-A-C-x_2) and a triangle (ABC). We know from geometry that the area of a triangle is found from $\frac{1}{2}$(base $\times$ height), so the area under the region x_1-A-B-C-x_2 is:

$$A = (y_1 - 0)(x_2 - x_1) + \frac{(y_2 - y_1)(x_2 - x_1)}{2}$$

But what if the curve is not so regular? Most of the curves derived from scientific experimental data are not so well behaved as to be straight lines or other regular geometric figures. For example, consider the curve in Figure 5-1b. We can still use rectangles to find the area under the curve, but we have to construct a large number of small rectangles, calculate and then sum up their individual areas.

Few of the rectangles will exactly contain the line $y = f(x)$. Some rectangles are oversized, while others are undersized. If there are enough rectangles, and they are each small enough, then those rectangles that tend to underestimate the area under that region of the curve are usually balanced by those that tend to overestimate the area. What happens as the width of the rectangles goes to zero ($x_2 - x_1 \rightarrow 0$) is one of the central tenets of calculus, and will be discussed early in any competent course on the subject.

If we want to find the area under the curve $y = f(x)$ in Fig. 5-1b, we need to know that the area of any one rectangle is:

$$A = (y_{i+1} - y_i)(x_{i+1} - x_i)$$

Figure 5-1. Integration involves finding the area under a curve. (a) For a straight line, or regular geometric figure, the task is easy; (b) for complex shapes we must break the area into a large number of smaller, regular shapes.

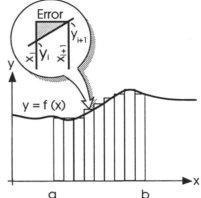

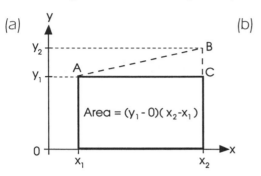

The total area under the curve from a to b is therefore:

$$A = \sum_{i=1}^{nx_i} (y_{i+1} - y_i)(x_{i+1} - x_i)$$

In the notation of calculus, the equation might be written:

$$A = \int_a^b f(x)\, dx$$

In some cases, you will want to find the average value of the curve. In that case, you will find the integral using an equation such as shown above, and then divide by the interval:

$$\bar{y} = \frac{1}{b-a} \int_a^b f(x)\, dx$$

In many cases, especially where the equation $f(x)$ is known, you would solve the problem analytically rather than graphically. But when the equation is not known, graphical methods come to the rescue. This topic is discussed in Chapter 13 when we deal with graphs.

Differentiation is the process for finding the *rate of change* of a curve, and the result is called the *derivative* of the curve. For a straight-line curve of the form $y = bx \pm a$, where b is the slope and a is the y-axis intercept, the derivative is merely the slope, b, which is rise over run:

$$b = \frac{y_2 - y_1}{x_2 - x_1} = \frac{\Delta y}{\Delta x}$$

When the curve is not a straight line, we can find the instantaneous derivative at any given point ("A" in Figure 5-2b) by taking the slope of a line s' tangent to the point. In this case, the term $\Delta y/\Delta x$ is usually replaced by the term dy/dx, which is the proper calculus notation.

(a)

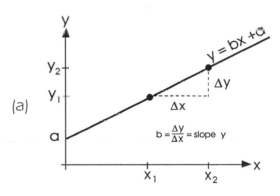

Conclusion

(b)

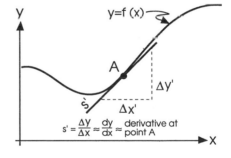

This chapter covered a number of basic arithmetic concepts that are founda-tions of doing science. In the next few chapters we will build on that foundation and erect some strong walls that allow you to hold—and understand— your experimental data. The first topic will be *averages*. Think you know what "aver-age" means? Don't be so sure until you read Chapter 6.

Figure 5-2. Differentiation is the task of finding the rate of change of a curve. (a) For a straight line of the form $y = bx \pm a$, the slope of the line $\Delta x/\Delta y$ is the deriva-tive; (b) for more complex shapes, the derivative is found at instantaneous points (such as "A") by finding the slope of a line tangent to the point (S').

6

What Is Average?

"**I**SN'T AVERAGE just, uhhh, average?" That's a common question, and the answer is not always so obvious as it might seem. There are several different kinds of "average," and all of them are valid in the right situations. The term refers to the most typical value, or most expected value, in a collection of numerical data. When you collect data, there are a number of ways that the results can vary from one observation to another (even when conditions are supposed to be the same).

First, of course, there is old-fashioned measurement and observational error. Not all rulers are truly the same, and not all applications of the same ruler to the same object turn out the same. Nor is it probable that even the same pair of perfect eyes will correctly read the scale every time a measurement is taken. In short, there will always be some variability in the measurements from one trial to another.

Next, there will be some actual variability in the events being recorded. Natural phenomena do, in fact, vary for one reason or another. One way to handle these variations is to find the most typical value for the lot. Consider the case where a horticulturalist observed a red berry bush over a period of time. At one point, the observer counted 28 bunches of berries, and found from 1 to 8 berries in the different bunches. What does "average" mean in this case? There are actually several different kinds of average, but the most commonly encountered are the arithmetic mean (usually called simply the *mean*), the *median*, and the *mode*. These are each a little different from the others, and all of them are correct "averages" when used in the right context. Let's look at the data a little more closely (see Table 6-1).

Bunch Number	Number of Berries
1	4
2	6
3	5
4	5
5	3
6	6
7	4
8	3
9	3
10	4
11	5
12	3
13	1
14	6
15	5
16	2
17	5
18	2
19	3
20	4
21	4
22	5
23	7
24	7
25	8
26	4
27	6
28	5

The arithmetic mean is the average most people use. The mean is nothing more than the sum of all values, divided by the number (n) of different values. Or, to put it in proper form:

$$\overline{x} = \frac{x_1 + x_2 + x_3 + \dots + x_n}{n}$$

The sum of all 28 values in Table 6-1 is 125, so what is the average?

$$\overline{x} = \frac{125}{28} = 4.46$$

The mean average is 4.46, although don't expect to find that "0.46" berry anyplace! This average is the arithmetic mean.

The median is another type of average: It is the middle value in the data set—that is, the value where exactly half of the values are above it and half are below it. In the present case, there are 28 values, which is an even number, so the median will be midway between two

Table 6-1. Data table showing number of berries in each bunch.

of them...with 14 above and 14 below. Figure 6-1 shows the data distribution and is a crude kind of bar graph. Count the *x*'s in each category from one end to the middle, and then the other end. Note that there are 14 values between 0 and 4, and 14 values from 5 to 9. That means the median value will be halfway between 4 and 5, or 4.5. If there were an odd number of data points, then the middle point—the median—will be the actual data point that has an equal number of points above it and below it.

The mode is also an average and is the most frequently occurring value in the data set. In the data in Table 6-1, the mode is easily seen in the *x*-chart. There were more bunches with 5 berries than any other number, so that's the mode. So, now we have an arithmetic mean of 4.46, a median of 4.5, and a mode of 5... and they're all the *average* of the data set!

Different averages are used for different situations. If the data are perfectly symmetrical, then the mean, median, and mode are the same number. In

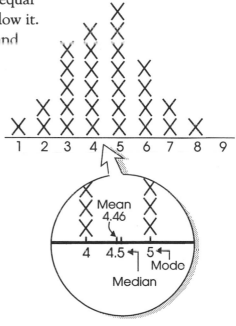

Figure 6-1. Three different "averages" (mean, median, and mode) and all of them are correct!

fact, that's nearly the case in the data above. If the mean, median, and mode are not the same, then the data is not symmetrical around the mean... and the difference is a test of that symmetry. In berry-bush data, the distribution is nearly symmetrical, so the mean could be used. But there are other situations where the mean is not terribly useful.

Consider small businessman George Thompson, who runs the *Big Barfburger Stop 'n' Shop*, and claims that the average salary there is more than $15,000 a year. But he also brags that he earned $60,000 there last year. He employs eight people, making a total of nine people on the payroll, including himself; two are full-time shift managers making $12,000 and $15,000. The other five are counter workers and short-order cooks, and make $8,000, except for two more experienced workers who make $9,000 and $10,000, respectively. These data are:

1.	Mr. Thompson	$60,000
2.	Senior Manager	15,000
3.	Junior Manager	12,000
4.	Cook 1	10,000
5.	Cook 2	9,000
6.	Worker 1	8,000
7.	Worker 2	8,000
8.	Worker 3	8,000
9.	Worker 4	8,000

That totals $137,000 annual payroll for nine people, or an average wage of $137,000/9 = $15,222.22. Notice anything odd? The "average" wage is more than the highest amount paid to anyone but Mr. Thompson! In this case the simple mean is misleading, and indeed that may have been Mr. Thompson's intent in citing the "more than $15,000" figure.

The median value is the value with half of the values above and half below (see Figure 6-2). In this case there are four people below $9,000 and four people above $9,000, so the median is $9,000 and the mode is $8,000. The median is the best "average" salary because the data are skewed towards one end of the range.

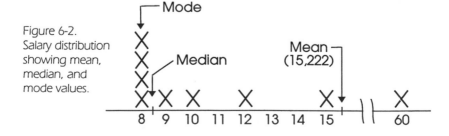

Figure 6-2.
Salary distribution showing mean, median, and mode values.

But what happens when we consider just the workers, and not the managers or Mr. Thompson? That means that we discard the $12,000, the $15,000 and the $60,000 data points, so:

$$\frac{8 + 8 + 8 + 8 + 9 + 10}{6} \text{ K\$} = \frac{51}{6} \text{ K\$} = 8.5 \text{ K\$}$$

("K$" means thousands of dollars.)

The mean is now $8,500, which is a bit more representative of the average wealth that the workers take out of the hamburger shop.

The mean is best used when the data are symmetrical, the median is used when the data are asymmetrical, and the mode is used to answer questions such as "What is the most common cause of death?" or "What is the most popular TV show on Friday night?"

A college professor I know has two sections of a freshman level physics course, and each section has around 100 students (the number accommodated by the lecture hall). Having nearly 200 students each semester makes it reasonable to keep certain statistics on the classes. When this professor gives an hour examination, which occurs five times each semester, the papers are graded by several graduate teaching assistants. A problem with using multiple graders is that, even in an "objective" monological subject like undergraduate physics, there is room for variation: There are easy graders, there are hard graders, and there is always at least one Attila the Hun in the group! In order to smooth out the data, this professor finds the mean for the entire group of 200, and plots the distribution of grades. If the mean and the median are more than 3 points apart (on a scale of 100), then the grades are adjusted. In order to keep some students from howling, the adjustment downward for the high scorers is considerably less than the adjustment upward for the low scorers (that's kindness operating, not statistics).

Similarly, the professor plots the distribution of the grades issued by each assistant. If the mean and median of the individual grade awards are out of kilter with the overall group, then an adjustment is made on the papers examined by that particular assistant.

Finally, because the professor has taught the same course for a number of years, he keeps statistical data on all previous times the class was offered. He then can compare each test this year with the same test in previous years, and weight the value of this test in the overall semester grade. He believes that an abnormal skew to the data means that either he didn't teach the material correctly, or the test was not truly representative of the material covered in the test period.

Interestingly enough, the distribution of letter grades follows a nearly bell-shaped curve even though it is believed by some colleagues that all of the adjusting makes him an "easy" professor. It turns out that the only people who are significantly affected by the policy are students whose raw grade fell close to a borderline. . . and in general they are helped by the policy.

What other averages are there? Well, there's the *geometric mean* and the *harmonic mean*. The geometric mean is used a lot when the data is not very symmetrical, especially in biological studies. Let's suppose that you have $48 to spend, and you spend one-half of your available bucks each day for 5 days. The data would tabulate like Table 6-2.

Table 6-2	DAY	AMOUNT
	1	$48
	2	24
	3	12
	4	6
	5	3

The arithmetic mean is:

$$\frac{48 + 24 + 12 + 6 + 3}{5} = \frac{93}{5} = 18.6$$

If we graph these values (Figure 6-3a), we note that the line connecting the tops of the bar graphs is not straight. To find the geometric mean, we need to find the *logarithm* of each value, add them up, and then take the *logarithmic mean*. Then we take the *antilog* of the log-mean. The log-mean is:

$$\frac{\log 48 + \log 24 + \log 12 + \log 6 + \log 3}{5}$$

$$= \frac{1.68 + 1.38 + 1.08 + 0.778 + 0.477}{5}$$

$$= \frac{5.395}{5} = 1.079$$

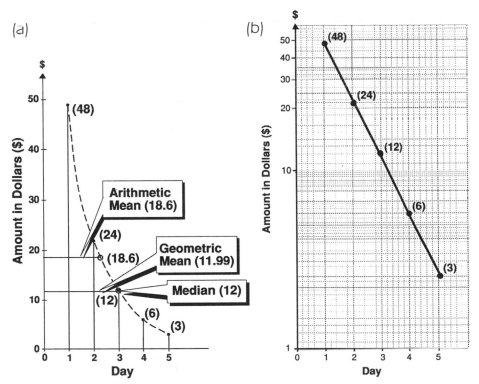

Figure 6-3. (a) Linear plot showing geometric mean, arithmetic mean, and median; (b) semilog plot of same data.

Now take the antilog of the answer...

$$\log^{-1}(1.079) = 11.99$$

The linear plot in Figure 6-3a is not a straight line. If we want to straighten out that line, we use *semilog paper* (Figure 6-3b). Keep in mind, however, that straightening out the curve is not the principal use of semilog paper.

The other mean, harmonic mean, is a bit more complicated, and is used when the data are expressed in *ratios*, such as miles per hour, or dollars per dozen, and so forth. The expression for harmonic mean reflects the fact that it is the reciprocal of the mean of the reciprocals of the data:

$$\text{H.M.} = \cfrac{1}{\cfrac{\dfrac{1}{x_1} + \dfrac{1}{x_2} + \dfrac{1}{x_3} + \dots + \dfrac{1}{x_n}}{n}}$$

Suppose we compare the price of eggs in the local store over one past month (See Table 6-3):

Table 6-3	WEEK	PRICE ($/Dozen)
	1	$2.29
	2	1.98
	3	1.56
	4	2.04

The arithmetic mean is:

$$\frac{\$2.29 + \$1.98 + \$1.56 + \$2.04}{4} = \frac{7.87}{4} = \$1.9675 \approx \$1.97$$

But the harmonic mean is:

$$\text{H.M.} = \cfrac{1}{\cfrac{\dfrac{1}{\$2.29} + \dfrac{1}{\$1.98} + \dfrac{1}{\$1.56} + \dfrac{1}{\$2.04}}{4}}$$

$$\text{H.M.} = \cfrac{1}{\left(\cfrac{0.437 + 0.505 + 0.641 + 0.490}{4}\right)}$$

$$= \frac{2.073}{4} = \frac{1}{0.518} = \$1.929 \approx \$1.93$$

Root Mean Square (rms) and Root Sum Squares (rss)

There are other "averages" that are sometimes seen in science and technology: *integrated average, root mean square* (rms), and *root sum squares* (rss). The integrated average is the area under the curve of the function (Figure 6-4), divided by the segment of the range over which the average is taken:

$$\bar{x} = \frac{1}{T} \int_{t_1}^{t_2} x \, dt$$

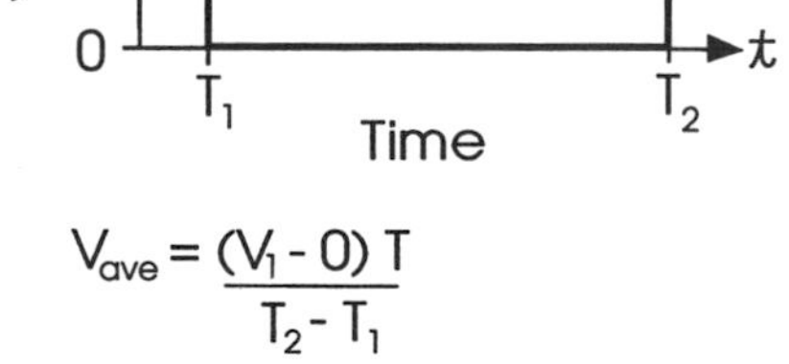

$$V_{ave} = \frac{(V_1 - 0)\, T}{T_2 - T_1}$$

Figure 6-4. Time average of a constant function is simple to calculate: it's the product of the height and duration.

The rms value is used extensively in electrical circuits and certain other technologies. For example, a sine wave alternating current (ac) wave may be compared with the direct current (dc) voltage level that will produce the same amount of heating in an electrical resistance. The value of the ac wave that is the dc heating equivalent is the *root mean square* (rms) value. The formal definition of rms is

$$V_{rms} = \sqrt{\frac{1}{T} \int_{t_1}^{t_2} V^2\, dt}$$

where T is the time interval t_1 to t_2. For the time being, you don't need to know how to evaluate this expression. That will come in a course on calculus.

For the special case of the sine wave, the rms value of voltage is $V_p/\sqrt{2}$, or $0.707V_p$, where V_p is the peak voltage (see Figure 6-5). For shapes other than sine waves, however, rms will evaluate differently.

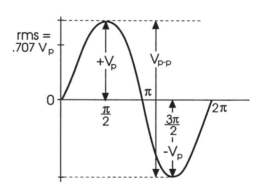

Figure 6-5. What's the average of a sine wave? Zero, actually, because the positive and negative values are equal but opposite, so cancel each other out. The root mean square (rms) value is more important.

The *root sum squares* is used in cases where different data are combined and they are in no way correlated with each other. For example, noise signals in electronic circuits are errors, and come from several different sources. Suppose we have n independent noise voltage sources (vn_1, vn_1, ... vn_n). When these sources are truly independent, they can't be simply combined in a linear manner, but must be combined using the rss method:

$$V_{rss} = \sqrt{\sum_{i=1}^{n} (vn_i)^2}$$

$$V_{rss} = \sqrt{(vn_1)^2 + (vn_2)^2 + (vn_3)^2 + \ldots + (vn_n)^2}$$

The rss method is used sometimes to define a single-valued error term from a number of unrelated error terms. Or, it may be used to find a single standard error of a number of measurements of the same value.

EXAMPLE

An electronic amplifier circuit contains five noise sources that produce the following unrelated noise signal voltage levels: vn_1 = 25 nanovolts (nV); vn_2 = 56 nV; vn_3 = –33 nV; vn_4 = =10 nV; and vn_5 = 62 nV. What is the rss value of a composite noise signal?

Solution:

$$V_{rss} = \sqrt{(vn_1)^2 + (vn_2)^2 + (vn_3)^2 + \ldots + (vn_n)^2} \ nV$$

$$= \sqrt{(25)^2 + (56)^2 + (-33)^2 + (-10)^2 + (62)^2} \ nV$$

$$= \sqrt{625 + 3136 + 1089 + 100 + 3844} \ nV$$

$$= \sqrt{8794} = 93.8 \ nV$$

Note well that the rss value is not the same as the summation of the components.

Conclusion

Although it is common in measurement, and in science experiments, to quote the "average" value of the number acquired, you must be cautious in using the correct average (or most reasonable average), and to correctly interpret what "average" means in the context of the experiment.

Experimenting—Step By Step

IT HAS BEEN SAID that all science is either physics or stamp collecting. (In fact, I said it in a classroom full of chemistry majors!) What this slightly insulting metaphor refers to is the two modes of collecting scientific data. *Directed experimentation* is one of the two principal modes used in science to uncover knowledge of nature (the other one being *informed observation*). In both cases, you are seeking informative events in nature about which conclusions can be drawn. When informed observation is used, the scientist goes to wherever such an event is likely to occur, equipped with tools appropriate to observing the expected event, and performs whatever actions are needed to increase the probability that an informative event will occur. In experimentation, the scientist contrives a situation, mostly in laboratories, that, as one professor put it, "increases the chance of an informative event occurring."

A scientific experiment is, therefore, a contrived event in which an investigator attempts to observe the effects of a *deliberate act*, performed in order to create an observable informative event. The word "contrived" should not be taken to mean in any way "false," but rather (according to a dictionary) "...a devise or plan...to bring about by stratagem...[to] create through ingenious means." Thus, an experiment is a stratagem, perhaps with an element of ingenuity, to create a deliberate act through which some effect or result can be observed.

All experiments are out to either prove or disprove something. Experiments are designed to prove something that actually happens and can be observed. It cannot prove a negative— i.e., the experiment cannot be designed to prove that something *won't* happen. No finite number of trials, no high number of repetitions of the experiment, can *ever* prove that it won't happen. After all, even millions of negative trials is not proof that on the *next* trial the result will not be positive. Gamblers tend to forget that fact a lot.

Also, no scientific experiment can either prove or disprove something that cannot be observed. Questions such as the beginning or end of the cosmos, or the existence of the spirit world, are beyond the realm of experimental science because such events are nonobservable. While the theorist can devise intellectual constructs of unobservable things, only those theories that have observable physical consequences are valid for scientific experimentation.

Before you can properly design a scientific experiment, it is necessary to understand what kinds of problems are researchable in the first place. So let's now turn our attention to determining what kind of problems are researchable.

Researchable Problems

Not every problem that can be conceived is actually researchable, and not every researchable problem lends itself to experimentation. Not knowing these differences can lead you to expend a large amount of time and effort on a largely unproductive activity. There are a number of reasons for lack of researchability in a problem. For example, the problem may not be practical to research because unambiguous data cannot be collected either at a reasonable cost, or at all. Second, the problem may not be truly practical for the question presently under consideration. You would not research the growth patterns of rice in Peking when studying the best material to use for road resurfacing in Virginia.

There are any number of other reasons why a problem is not researchable, but perhaps the most difficult to appreciate are the dread *j*-problems. (I'm indebted to author Don Lancaster for pointing out the existence of *j*-problems, although he called it the search for *j*-dollars.) Engineers use the lower-case letter *j* to denote the imaginary operator in mathematical expressions (denoted by *i* in other disciplines—both mean "the square root of minus one"), so one may conclude that a *j*-problem is an imaginary problem that seems real. Examples of *j*-problems include those which are misconstrued to be a problem, when in fact it is only that sufficiently accurate information is lacking that makes a situation seem like a problem. Such problems fall quickly when close scrutiny is applied.

Chief among *j*-problems, even in scientific circles, are those generated by rumor mills. A situation is noted and reported to others, but the original accurate description is distorted in the transmission from person to person. By the time the supposed "information" reaches the researcher, it is so garbled that the wrong problem—or even a nonproblem—is researched. Rumor mills exist in science, but we prefer to think that we are more sophisticated than others.

Another example of a *j*-problem is the "foggy mountain breakdown," that is, a problem that appears like a mighty mountain, but is so shrouded in the fog of misperceptions that no real understanding is possible. When the problem is shrouded in fog, you can be certain that the research effort will break down.

First Law

All of which leads us to the first law of defining a researchable problem: *The problem must be clearly formulated in words.*

If the problem cannot be defined properly in words, preferably in writing, then it is a sure bet that the solution will remain elusive. Without clear definition, you will either fritz around without direction ("Brownian Motion" school of research), go off on an irrelevant tangent ("Lost in Space" school of research), or go around in circles ("Chase Your Tail" school of research).

What constitutes a properly defined problem? Some people are tempted to write a novel-length specification of the problem. Unfortunately, that practice often complicates rather than clarifies the problem. A book editor from a major publishing house tells hopeful writers that the longest book, or even series of books, must have an *single underlying theme* that can be summed up in a single sentence of 25 words or less. If the main theme cannot be so summarized, then the project is not well thought out. So it is with scientific problems: If the problem cannot be succinctly stated in a few words, a single paragraph at most, then it is probably either too global, too ambiguous, or it has not been broken down into small-enough pieces.

Another element that makes a problem nonresearchable is a lack of testability. Before a problem can be researched it must be of such a nature that it is testable through empirical evidence. In other words, it must be such that hard data can be generated by either experience, observation, or experimentation. To the maximum extent possible, the stated problem must be *quantifiable* so that the data is numerical, and thus lends itself to the discipline of statistics. Opinions, "gut feelings," and other subjective factors have their place, but not in a scientific experiment.

Second Law

The second law of researchability is: *The problem must be such that useful empirical data can be identified and collected; and wherever possible that data should be numerical in nature.*

The emphasis on numerical data is not because of any mindless prejudice against subjective information, but, rather, properly taken numbers are less subject to misinterpretation and opinion, and they allow for the use of statistics in the analysis.

Finally, the problem must be such that it is feasible to collect data through either direct or indirect means. If data cannot be collected, then it is a waste of time to attempt the experiment. There must be some *observable* criteria on which to base an experiment, or else the whole enterprise is little more than guessing, or sheer mysticism. And "observable" could mean that it is accessible. If you can't physically observe the data, either with the senses or with tools such as instruments, then there is little likelihood short of mystical revelation that it will be of use.

Third Law

The third law of researchability is: *The events on which data are collected must be accessible and observable, either directly or indirectly, through whatever means are appropriate for the case at hand.* Deciding the type and quantity of data to collect will result from properly defining the problem in words, and then designing the experiment to reach the truth.

Stepping Through Research

There is an unfortunate tendency for some people to look to the "Eureka Experience" to advance their scientific careers. People tend to look for the brilliant flash of insight, the breakthrough, or the sudden revelation or "quick solution." But that is not the usual stuff of scientific progress. Does it happen? Of

course it does, but it's rare. . . and even when it does occur it must be validated by proper observation or experimentation. A structured approach to doing science follows a logical, step-by-step procedure. While one could argue that other methods also pan out, the logical step-by-step method reliably leads to success more often than the Brownian Motion (random walk) approach.

Step 1: Clearly define the problem. If the problem cannot be clearly defined in writing (which forces you to verbalize it), then you basically haven't got a researchable problem. The problem statement should express a point of view on what is or is not true, and then propose an experimental path to solution. Furthermore, the problem statement must more or less obey the three laws discussed above.

Step 2: Describe the design of the experiment in writing. The "design" of the experiment consists of two different types of activities. You may be tempted to think in terms of the most visible: The configuration of apparatus (or whatever) that will be used to perform the experiment. That is really a matter of data collection method (see below), and is the naive design. The other connotation of "design" is more philosophical in that it sets the approach to doing the experiment. Is it a factorial experiment? Is it a simple comparison experiment? Is it a matched comparison experiment? Is it a sequential experiment? Or is it some combination of these? (We'll discuss the various types of experiments later in this chapter.)

By the way, the time to consult a competent statistician who can help you with the design of the experiment (which is never a bad idea) is prior to doing the experiment or even committing any significant resources to it. More papers are rejected, and more research grants are denied, because of one single factor than any other: Poor design that tells the reviewers "Ya can't get there from here!"

When the experiment is completed, and the data recorded and ready for analysis, it may be too late for the statistician to help. All too often, researchers come to the statistician for their calculation function *post facto* without realizing that their more valuable efforts are before the experiment is performed. Indeed, with a simple calculator, or if appropriate, a PC-based statistics software program, you can do the calculation work yourself. But the statistician can help you (a) design the experiment to eliminate problems, and (b) interpret the results to ensure that the right conclusion is drawn.

If your statistician colleague always has a pained, almost grieved, look on her face when you come down the hallway, arms burdened down by a quarter-ton of laboratory notebooks, computer printouts, and scrap note paper stained with midnight coffee or other noxious stuff, then she might be trying to tell you something. Statisticians are the misunderstood profession. They are of immense use, but at the opposite end of the experimental project than where they are normally consulted: *At the beginning.*

Step 3: Describe the method of data collection in writing. This statement should cover the experimental configuration, the instruments to be used (if any), the nature of the data to be collected, accept/reject criteria where appropriate, and the format in which it will be recorded. Also any indicators for decision points such as when to proceed or quit the experiment, or when to take an alternate direction should be included. A step-by-step procedure for actually carrying out the experiment is also part of this effort. It is here that the "turn the big red knob and read meter M1" type of instructions are set forth. It is important, especially in repetitions of the same procedure, that it be done the same way each time.

Step 4: Perform the experiment and collect data. Carry out the experiment, being very careful to closely adhere to the scheme decided on in the previous step. Note any irregularities, and keep good records.

Step 5: Analyze the resulting data. Analysis of the data is done to draw conclusions on the results of the experiment. Was the hypothesis accepted or rejected? Infer whatever significance is warranted by the evidence, and no more. If the problem was properly stated, the experiment was properly designed and carried out, and the data is not corrupted by some external factor, then the analysis and interpretation should be relatively easy (*Oh yeah?*). I cover this important topic in later chapters.

But what if the data are not easy to analyze or interpret? Then consider the possibility that one or more of three things went awry: Either there was (1) an unaccounted experimental artifact or error; (2) there was some variable not properly isolated (we either did not ask the right question of nature, or the question that we asked was insufficiently isolated); (3) the design of the experiment was flawed from the beginning.

FAILED EXPERIMENTS

All experimenters should remember the words of Professor Julius Sumner Miller, a leading teacher of science on TV a generation ago. Experiments NEVER fail, experimenters do. Nature does what nature must. If your experiment seemed to fail, then look not to the experiment per se, but rather to its design or implementation.

Step 6: Submit to others for criticism. Science does not make progress in isolation, but rather through peer review, a vigorous interchange of viewpoints and experimental results, and mutual criticism. It's painful, I know, but research must be submitted to others for critical review before we can claim that the experiment is completed... for it might not be. There may be snakes with long critical fangs lurking in the grass.

Forming and Testing the Hypothesis

The basis for any scientific experiment is a *hypothesis* about the phenomena being investigated. According to one dictionary, a hypothesis is the statement of a "...tentative assumption made in order to draw out and test the logical or empirical consequences" of a situation regarding some population under study. The hypothesis (designated H_1) may or may not be true, but hopefully the experiment will either accept the hypothesis (i.e., show that it is true), or reject ("falsify" or show that it is false). For example, we may hypothesize that "A occurs more frequently than B," and then create an experiment to find out if that statement is true or false.

A *null hypothesis* (designated H_0) is a statistical proposition to be tested and either accepted or rejected in favor of an alternative, such as the probability of occurring by chance. In the null hypothesis we tentatively assume a negative outcome. If it does occur, then the null hypothesis is clearly invalidated. For example, suppose you want to find out if kerosene can be ignited by a lighted match tossed into an open dish of the fluid. If you toss a lighted match into the kerosene and it is doused, then you cannot claim that "kerosene is not ignited by tossed matches" no matter how often the experiment is repeated. (DON'T ACTUALLY TRY IT—IT WILL IGNITE AND BURN!!!) But if the hypothesis is that "kerosene will NOT ignite when a match is tossed into it," then the very first trial is likely to invalidate the null hypothesis in a roaring flame.

The hypothesis is the basis for the experiment. For example, "in growing tomatoes, Brand-X fertilizer is better than Brand-Y fertilizer." We could then grow two or more sets of tomatoes under the same, or very similar, situations, and decide whether the hypothesis is true or false. First, of course, we must define what "better" means. If more pounds of tomatoes were produced under the influence of Brand-X than Brand-Y, and "better" in this case means "more pounds," then we might reasonably accept the hypothesis, and assert that it is true. Alternatively, if we find that Brand-Y produces more pounds of tomatoes, then we reject the hypothesis and assert that Brand-X is not superior to Brand-Y.

A good practice in experimentation is to form a null hypothesis (H_0) that you hope to reject, the inverse of the hypothesis that you hope to prove. If you succeed in rejecting H_0, then you implicitly accept its inverse, which we'll designate H_1. Consider a medical example. A researcher wants to compare two different treatments for the same disease. The old treatment involved surgery, and was known to be 25% effective. In this case, "effective" means the cure remains robust—i.e., does not reoccur—after a predetermined period of time (typically 5 years for cancers). The 25% figure for surgery is well-established historically on the basis of many years of use. The experiment is designed to test a new treatment using a drug (D) against the old surgical treatment (S).

We can state that historically the probability of success with S is:

$$P(S) \ = \ \frac{25}{100} \ = \ \frac{1}{4} \ = \ 0.25$$

(See Chapter 9 on probability.)

If $P(D)$ after the experiment is significantly higher than 0.25, then we may conclude that D is superior to S. Our null hypothesis is: "D is not superior to S."

Experiments as Comparisons

Experiments are basically comparisons. Two events may be compared with each other, or an event may be compared with the probability of obtaining the same result through the operation of chance alone. For example, you might compare the action of two different drugs in treating similar patients, or compare the effectiveness of a new clinical laboratory test procedure against the probability of correctly guessing the result (by chance).

In some experiments, "naive chance" is used. Naive chance predicts that the next occurrence of an event will be exactly like the last occurrence. For example, the ability of a method, or theoretical model, to predict something like the weather or the daily performance of the stock market is based on a comparison of the new way and the "naive" way in which tomorrow's performance is predicted to be identical to today's performance. Thus, tomorrow's weather is predicted to be identical to today's weather; if it rained today then we predict that it will rain tomorrow. Alternatively, if the stock market Dow-Jones Industrial Average changed +50 points today, then it will also change +50 points tomorrow. We get some idea of the validity of our hypothesis when two different prediction methods are both compared with the naive model.

Don't try to get rich using the naive model to predict stock or commodity prices, or the roll of dice... however, you might beat some of the self-appointed "experts" who solicit big bucks from non-professional investors by mail.

In the following sections, we'll look at several types of scientific experiment.

> **WANNA GET RICH?**
>
> A number of "methods" for picking stocks are offered by financial consultants, planners, and just plain charlatans. When compared with the naive model, none of them to date beats the method traditionally used by a large number of institutional investors: Buy a large, diverse portfolio of the Standard & Poors 500 (at least 100 different issues), and depend on random chance. The performance typically beats the "methods" by several percentage points when measured over the long term.

Paired Comparisons

In this type of experiment, two or more events are compared to each other, not against pure chance. There are three basic forms of paired comparison: (1) simple matching; (2) symmetrical matching; (3) split samples. (This is from *Practical Statistics*, by Russell Langley.)

In the simple matching type of paired comparison, two samples are compared with each other after being randomly selected from the same pool of the larger population. For example, from a large pool of patients suffering headaches, you could randomly select some for the control group and others for the test group. When a pain reliever is administered, the results of the two groups are compared with each other. While you must guard against inadvertent bias (or in the case of some advertised claims for such remedies, perhaps not so inadvertent), the method yields significant data on the relative merits of two different things. Notice the word "relative," for this method says nothing about any absolute merits.

In a symmetrical matching test, there may be a large number of subjects, but each one is judged only against itself, but with part treated one way and part treated another way. At several points around the country you can see short sections of multi-colored highway pavement. These are placed by highway engineers to test various types of road surface. If two 20-ft long sections of different materials are placed sequentially to each other, then all of the traffic that flows across one will also flow across the other at the same speed, and both are subjected to the same environmental conditions. The relative characteristics of the two materials can then be compared not only with each other, but also with the normal concrete or asphalt surfaces at either end of the test strip.

Similarly, in medicine a test treatment might be tried on one side of a sunburned back, and another treatment on the other side. In toxicology tests, it is common to embed new materials in a test animal's subcutaneous tissue along with a known substance at another location close by. The goal is to compare the reactions of the tissue to the two different materials, and produce a relative assessment of the toxicity of the new material.

A potential source of error is any unaccounted interaction between two factors being compared. For example, two metallic materials close by each other in living tissue may cause damage unrelated to their respective toxicities because of ionic electrical currents between them that would not exist if only one material were implanted.

Split samples are used in some cases to see how a single thing reacts under two different sets of circumstances. For example, chemicals and metals can be broken into two or more lots and tested under different protocols. A paint manufacturer may take a sample from a vat of paint and test its storability. The sample could be divided into three lots: One for a colder than normal environment, one for a hotter than normal environment, and one to be stored in the accepted or recommended environment. From an evaluation of the effects on the three lots from the same vat, you can infer something about the storability of that particular formula of paint.

An example of a split sample can be seen in a 24-bed hospital Post Coronary Care Unit (PCCU) where the patients are sent to further recover from a heart attack, after release from the intensive Coronary Care Unit (CCU). Each patient in PCCU wears a miniature radio transmitter that telemeters their electrocardiograph (ECG) signal to the central nurses station where it is monitored on an oscilloscope and by computer. The head nurse complained that she buys the 9-volt "transistor radio" batteries that operate the transmitters in case-lots of 10 batteries to a box, and 20 boxes to a case (200 batteries per case). The telemetry system manufacturer led her to believe that a battery should last 24 hours, but she found that to be true only shortly after the batteries were delivered. After a while, as the batteries aged, the useful life deteriorated so much that a battery change was needed once on each 8-hour nurses' shift.

After a biomedical equipment technician suggested that the life is extended by refrigeration, she arranged a little experiment. The next case to arrive was split into two samples: Half were placed in the normal storage closet at the nurse's station, and half were placed in an ordinary household refrigerator (the same one used to hold units of blood, drugs requiring refrigeration, the nursing staff's bagged lunches, old pizzas, and other gooey stuff. . . not all of which was legal to store together). A protocol was arranged whereby fresh replacement batteries were drawn from either source on a random basis. Results on the first batch tested suggested that the refrigerated batteries lasted longer "on the shelf" than those stored at room temperature. After testing about ten cases, over a period of several months, and constantly monitoring the temperature both in the closet and the 'fridge, she determined that the technician was, indeed, correct.

Double Blind Experiments

In a double blind experiment, both the application of the trial method and the evaluation of the results are blind. For example, in medicine a new drug is usually tested in a double blind protocol. The patients are randomly divided into two groups, a test group and a control group. The test group receives the new drug, while the control group receives a substitute that is inert. Neither the physician who evaluates the effects, nor the nurse who administers the drug, will know which patient received which substance. In some tests, three different groups may be randomly created. One will receive the new treatment, one will receive an inert treatment, and the third group receives no treatment. This protocol is followed to overcome the placebo effect.

A problem in medical experiments is that professional ethics forbids a true double blind protocol in many situations. A doctor can't ethically deny a potentially curable patient a known good treatment in the interest of scientific rigor. As a result, many medical treatments are tested under less than optimal conditions, especially when the new treatment works best in the early stages of the disease.

Factorial Experiments

Although it is often said that the goal of the experimenter is to control all variables except the thing being studied, that is not always possible in the real world. Indeed, it might not even be desirable in many situations, especially where several factors are normally present and are interactive. A *factorial experiment* is one in which two or more factors are examined in a single experiment. The factors can be studied for the effects when they are absent and present, or at two or more different concentrations, or any of a host of other differences. Instead of setting up separate situations, which always introduces an element of ambiguity into any experiment, we instead seek to perform a single experiment without losing any relevant data. Factorial experiments are often used in agriculture, medicine, and other life sciences.

Let's take as an example the problem of determining which fertilizers work best in growing tomatoes in a greenhouse. We know there are two basic forms of fertilizer, which we will call Brand-X and Brand-Y. If a blind design is adopted, then neither the person who administers the fertilizers, nor the person who weighs the crop and evaluates success will know which fertilizer was used on which plot; only the plants will know. In addition, quantities and concentrations must be rigorously controlled so that some inadvertent misadministration doesn't bias the results.

One way to study the relative merits of these two fertilizers is to break our growing area into three plots such that one area uses x, a second area uses y, and the third area uses a combination of both types of fertilizer in equal amounts (xy, which signifies $x+y$). Each plot is started with replanted baby tomato plants grown from the same batch of seeds. The plantlings assigned to each plot are selected at random. All plots are made of the same amount of the same soil. The idea is to keep every-thing as common as possible between the plots, except the two fertilizers. The index of success is the total weight of tomatoes obtained from each plot.

The plots can be divided as shown in Figure 7-1a: All three plots side-by-side so that all are affected equally by environmental effects such as humidity, temperature, and light. Alternatively, if one or more of these effects is not easy to control in any one spot (light and temperature can vary markedly from place to place, even in a small structure), then you might opt for a layout such as Figure 7-1b.

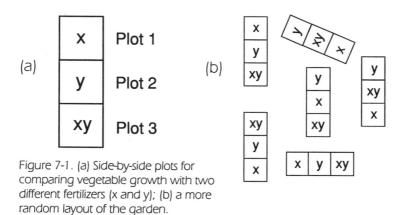

Figure 7-1. (a) Side-by-side plots for comparing vegetable growth with two different fertilizers (x and y); (b) a more random layout of the garden.

Suppose our structure is per Figure 7-1a, and the following results are obtained:

$$x = 50 \text{ lbs of tomatoes}$$
$$y = 40 \text{ lbs of tomatoes}$$
$$xy = 60 \text{ lbs of tomatoes}$$

It is clear that Brand-X produced more tomatoes than Brand-Y, and that the combination of x and y produced more than either alone. On the surface of it, we may conclude that mixing equal portions of x and y is the best way to increase our yield of tomatoes. But that assumption is not valid, for we don't know whether the effect was *additive*— the combined effects are the sum of the individual effects, or *interactive*—the combined effects are either more or less than the simple summation of the two components. Similarly, we don't know what the effects of using no fertilizer are on the yield, so we can't legitimately make any assumptions about the merits of *x*, *y* and *xy*. You have to guard against drawing conclusions without taking into account the possibility of *synergism*, an interaction such that the whole acting in concert is more than the sum of the individual effects. Such effects might actually increase the yield, or reduce it.

A better experimental method, one that allows for testing interaction between methods, is seen in Figure 7-2. In this layout, there are four plots in the greenhouse: One is treated with Brand-X; one is treated with Brand-Y; another is treated with equal amounts of Brands *x* and *y*; while a fourth is treated with none (*n*). Again, every factor other than the fertilizers is kept constant, so we can judge the effects of the variables. At the end of the growing season the following totals are recorded in the log book:

$$x = 50 \text{ lbs of tomatoes}$$
$$y = 40 \text{ lbs of tomatoes}$$
$$xy = 60 \text{ lbs of tomatoes}$$
$$n = 25 \text{ lbs of tomatoes}$$

By using four plots, we gain the ability to more accurately judge the effects of the different treatments. We can examine the equality:

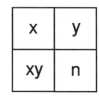

Figure 7-2. If each cell is laid out in this manner, a lot more data can be calculated.

$$xy = n + (x - n) + (y - n)$$

If this equation is balanced (i.e., both sides equal), then there is an interaction present between factors *x* and *y*; inequality implies that there is no interaction between *x* and *y*. Using the data from above:

$$60 = 25 + (50 - 25) + (40 - 25)$$

$$60 \neq 65$$

We can therefore conclude that there is no interaction between the two fertilizers, and that the effect is simply additive.

We can use the same data to examine other possibilities. We know that the effect of x in the presence of y is:

$$xy - y$$

...the effect of x in the absence of y:

$$x - n$$

...and an indication of interaction:

$$I_{xy} = (xy - y) - (x - n)$$

The factorial design shown above for two factors can be extended to a number of factors. For example, you could set up plots for heavy application of x, light application of x, heavy application of y, and light application of y; these could be designated x_H, x_L, y_H, y_L for purposes of analysis.

Sequential Design

A surgeon was planning an experiment in which a new implantable device was to be tested. The device is quite expensive, and the operation is quite extensive. If a predetermined (and fixed) number of patients were operated on, then the program would be both very expensive and quite lengthy. The surgeon's statistician recommended a *sequential design* in which the results are continuously monitored so that the experiment can be terminated when a predetermined level of significance (or nonsignificance, for that matter) is obtained.

The sequential form of experiment was invented by a mathematician named Abraham Wald, who was a professor at Columbia University in New York City. Wald was 36 years old in 1938 when he emigrated from Nazi-occupied Austria to the United States. During World War II, Wald headed the Statistical Research Group at Columbia. The sequential method was recognized as so significant that the United States Government classified the method ("Restricted"), and then used it in more than 6,000 war goods factories, where it allowed them to get product "out the door" faster. In 1945, the method was released to the public.

Experimental Problems

A problem that plagues a lot of experiments is a lack of focus. The design of the experiment should be such that it leads to clear and unambiguous results. One of the roots of controversy in many scientific efforts is that the investigators have not yet asked the correct question of nature. It reminds me of the vitamin-C squabble. For decades the question has remained unanswered: Does it or does it not prevent the common cold? If you look at the literature, it seems that some studies confirm one hypothesis, while others confirm the opposite hypothesis. Except for the fact that this particular controversy seems overburdened with investigators who strongly hold previous opinions on the matter, one would suggest that the ambiguity may arise from a lack of focus: Have the investigators really isolated the factors and controlled all except the unknown variable?

If the experiment does not yield clear, unambiguous results, then there will be a perpetual question mark hanging over the conclusions drawn. Of course, the unethical scientist might point out that ambiguity is beneficial for those whose work is supported by annual grants. The old maxim

> **WILL DR. MURPHY PLEASE CALL DR. FINAGLE?**
>
> Murphy's Law: If anything can go wrong, it will.
>
> Helms's corollary: Murphy was an optimist, especially in New Hampshire.
>
> Carr's corollary: If anything can't go wrong, it will.
>
> Finagle's Law: (Murphy's Law if you went to grad school) The perversity of the universe tends to a local maxima whenever and wherever it can do the most damage to your experiment

goes something like this: Always raise more questions than you answer through your research, or the next year you are out of business. In other words, when all is settled, your job is finished…as is your paycheck.

There are many situations where it is not possible to control all of the variables, or even know in advance what the variables are. In some rare situations, you might not even know that other variables exist, despite the fact that they affected the outcome of the experiment. Such effects can be seen in all of the sciences, but perhaps in the social sciences it is most common.

A way to overcome these effects is to look at the same problem from different perspectives. You might perform more than one kind of experiment on the same concept in order to see if the results are consistent. If there is an inconsistency, then you may reasonably suspect that something is amiss. For example, one or both experimental theories might be wrong, or at least substantially so. Or there may be a problem with the measurement apparatus; instrument calibration errors could lead you to the wrong conclusions. In some cases of normally ambiguous experiments, a scientist may apply multiple techniques to find the narrowly focused truth that eludes any single method.

Interpretation

Once the data are collected and analyzed (discussed in detail in following chapters), it is time to figure out what they mean. In other words, an interpretation of the significance of the results of the experiment must be provided. It is here that we sometimes see problems, especially where the topic is controversial or where there is a significant emotional content. While these problems surface frequently in the social sciences, they are also not exactly unknown in physics, chemistry, engineering, and biology. Overenthusiasm and hyperego are frequent causes of improper interpretations of experiments. They may be different causes, but they both nonetheless lead to the same error: More is claimed for the experiment than can be justified by the data.

That last comment is so important that it bears repeating, and enshrinement as a Rule of Thumb:

Do not make more claims than the data can support (leave that for attorneys).

An example can be seen in the widespread reporting in 1991 of a study in which supposedly matched pairs of black and white men were sent out looking for work. It was proved that the blacks ran into more racial discrimination than the whites, but the reportage of how much drew a lot of fire. While a co-author of the study, appearing on a radio talk show, did not make excessive claims, the radio audience and host seemed to make very wild claims that the discrimination was tremendous—black males faced considerably more difficulty than equally qualified whites. It was stated by some commentators that whites were successful three times more often than blacks. While some critics have faulted the experimental methods used, let's assume that the investigators properly executed a reasonably well-designed experiment.

The reported results were these: White applicants advanced further than blacks 20% of the time, while blacks advanced further than whites 7% of the time. Commentators claimed that the data indicates that blacks were favored one-third as many times (7/20) as whites, and therefore whites are three times more likely to advance further in the hiring process than blacks. What was overlooked is that *blacks and whites advanced equally as far in 73 percent (i.e., 100 − 20 − 7) of the cases.* Thus the two actual preferences are:

Whites: 73 + 20 = 93%
Blacks: 73 + 7 = 80%

The advantage of whites in this test is not 20/7, but 93/80, which computes to 1.1625, or an advantage of a bit over 16%. The correct interpretation of the data is not that massive racial discrimination exists in the job market, but that *some* exists... and even then the results are suspect until one knows more details about how the experiment was conducted, and replication confirms it.

Other commentators were just as erroneous in claiming the opposite conclusion: That the data showed that tremendous progress had been made in race relations, and that little racial discrimination exists in the job market. Nothing of the sort is shown by that experiment, for we don't have a similarly structured previous experiment on which to base a comparison.

You can make *guesses* of factors outside the scope of the data and the experiment that generated it, but these must not be claimed as results. Rather, discuss them in terms of possibilities and conjectures that call for additional study—and clearly label such material as speculative, and do not claim it as fact that is proved by the data.

Presentation of Results

There is a cartoon that shows a ramshackle outhouse, complete with crescent moon carved into the door. The caption says: "The job is never done until the paper work is finished." After an experiment is performed, the data collected and analyzed, and an interpretation made, the next part of the process is to present the results to others. In a professional setting, that means publication in a scientific journal or the printed proceedings of a formal conference. In lesser settings it may mean a report or oral presentation to a teacher, a boss, or to friends and colleagues. Presentation is the process by which the results become known to others, and criticism is invited.

Criticism *invited*? Are you kidding? Yes, criticism should be invited. Science makes progress only occasionally through giant breakthrough leaps into a whole new paradigm; most scientific progress comes almost quasistatically as each researcher in turn builds upon the previous work of either themselves or others. Presenting the experiment is actually the process of obtaining feedback to help improve the knowledge of the subject being studied. Others will make suggestions, or point out flaws, and you (or someone else) will be able to (1) replicate your experiment and (2) build on your results to deepen their own understanding of that phenomenon or some related phenomenon.

When everyone submits to the same discipline, a synergy builds that raises the knowledge of everyone involved...unless egos and emotion become involved (and they never should). I am convinced by group dynamics experts, and my own observations, that there is a powerful amplification factor when several people "brainstorm" a problem, as I discussed back in Chapter 2. The collective output of several brilliant minds can be devastatingly greater than the output of a single mind... the whole is greater than the sum of the parts, which is what synergy is all about.

The idea of replicating experiments is crucial because it permits others to follow your method and discover whether or not it leads to the same result. Very often, you discover uncontrolled factors that only exist in a particular laboratory, and these can pollute data. In one case, there was some undiscovered electrical coupling between two sets of signal wires that was causing erroneous but believable signals to be generated. The laboratory was used to study *evoked potentials* in electroencephalograph (EEG) waves picked up from the scalp through electrodes. By flashing a light repetitively and then coherently averaging the EEG signal taken from scalp electrodes placed 2 cm either side of the cranial structure called the *inion*, it is possible to dramatically reduce all components of the acquired electrical signals, *except* those due to the optic nerve being stimulated by the light flashes.

Unfortunately, some results were produced that other labs could not replicate (even in the same building!). The scientist who owned the EEG lab was troubled by nonreplication, so investigated and discovered that the stimulating electrical trigger pulse signal sent from the computer to the flashgun was coupling into the wires from the electrodes on the victim's (errr..."subject's") scalp, and were treated by the computer as valid EEG signals...and they coherently integrated with the flash because they were related to it in time. Embarrassing, perhaps, but that is how progress is made, and errors exposed.

Although there are many different methods of presenting experimental results, there are a few common elements that are considered good practice.

> **WARNING: GRADUATE STUDENTS CAN BE HAZARDOUS TO YOUR HEALTH**
>
> Anyone who works around physiology departments in medical schools, or certain other scientific venues, know one thing for certain: Stay to heck away from graduate students who are nearing their thesis research experiments! They need victims for the experiment, and are in the habit of shanghaiing anyone who comes along, even the department chairman. Thanks, Nancy... (you know what icey things you did to our forearms in Ross Hall!)

1. Provide a cogent statement of the problem being investigated. This statement should be written as clearly, and be as well focused as possible.

2. Discuss the literature search results: Who in the past has investigated this phenomenon, or related phenomenon, and what were the results they found? Also discuss any criticism of those results that were published later. On certain current topics (where a lot of current work is known) go back in the scientific literature at least three years, and on other topics go back at least ten years.

3. Precisely describe the method used in your experiment. Include a discussion of any apparatus or instruments used in making the measurement. Diagrams are nice, when appropriate.

4. Describe the actual performance of the experiment, noting any unusual occurrences or anomalies that were experienced. Remember Milligan's Law here para-phrased, When something looks funny, Record Amount of Funny.
5. Tabulate the data and calculate the relevant statistics (for example, perhaps the mean, variance, standard deviation, among others). Give all raw data, unless it is too cumbersome. It is often assumed (rightly or wrongly) that an experimenter who refuses to divulge raw data has something to hide—and, after the disclosure of too many scandals, we know that sometimes the charge is correct.
6. Discuss any negative results. It is patently unfair, and bad science, to only present those facts that support the hypothesis or conclusions drawn by the researcher. If contrary evidence surfaced, then it must be presented and, to the extent possible, explained.
7. Present any further questions that arose in the course of the study. These questions may well spur further research on your part, or on the part of others.
8. Include every factor that will aid other researchers in replicating the experiment.

Causality or Mere Correlation?

The fact that B follows A does not necessarily imply that A causes B. For example, we can show rather conclusively that the number of births in Los Angeles between 1918 and 1960 is highly correlated to the concentration of atmospheric pollutants over the same period. Are we to infer that smog causes fertility, or that both automotive emissions and births increased because of increases in population? Also, have you noted that yellow traffic lights precede red lights? Do we therefore assume that yellow lights cause red lights?

The error of assuming "A causes B" because A precedes B is sometimes called the *post hoc* fallacy (see Chapter 4). John Tukey and Frederick Mosteller in *Data Analysis and Regression* propose three conditions that must be satisfied before a causal relationship can be established: (1) consistency, (2) responsiveness, and (3) mechanism.

The consistency requirement means that the two correlated factors are routinely found together. If a test is run several times, and the same sort of correlation (if not exactly the same result) turns up in these subsequent trials, then we may assume that the result is consistent. Responsiveness means that a change in the independent variable results in a change in the dependent variable. This can be observed in experimental settings, and should be tested. Finally, a mechanism must be established. We can correlate many things that are both consistent and responsive, but unless there is at least a viable theory on how the two factors are related, then all that we can claim is a correlation between them.

The word "causality" is often used to mean situations where a strong correlation exists, but that is a statistical usage of the word rather than a philosophical distinction, and should not be taken too literally. "A causes B" in a formal logical sense does not imply that A is literally caused by B, unless you can show consistency, responsiveness, and mechanism; rather "A causes B" tells us that a correlation of some sort exists.

In Summary

That's experimentation in a nutshell. In the next few chapters, we'll take a more detailed look at some of the topics relevant to good experimenting.

References

George Box and Soren Bisgaard, "The Scientific Context of Quality Improvement," Center for Quality and Productivity Improvement, University of Wisconsin-Madison. Report No. 25.

A.J. Jaffe and Herbert F. Spirer, *Misused Statistics: Straight Talk About Twisted Numbers*; Marcel Dekker, Inc. (New York & Basel, Switzerland, 1987).

Russell Langley, *Practical Statistics*, Dover Publications (New York, 1968, 1970). Reprint of a British book.

John Tukey and Frederick Mosteller, *Data Analysis and Regression*, Addison-Welsey (Reading, MA, 1977).

Taking Measurements

MEASUREMENTS are made to fulfil one or more of several different goals: Obtain information about a physical phenomenon, assign a value to some fundamental constant, record trends, control some process, or correlate behavior with other parameters in order to obtain insight into their relationships. The physical quantity being measured is called the *measurand*.

There are three general categories of measurement: *Direct, indirect,* and *null*.

Direct measurements are made by holding the measurand up to some calibrated standard and comparing the two. A good example is the ruler used to cut a piece of lumber to the correct length. You know that a 2 x 4 must be cut to a length of 24 inches, so you hold a yardstick (the standard of reference) up

"Looks like another long year..."

THE YEAR

to the piece of lumber (See Figure 8-1), set the zero inches point at the butt end of the 2 x 4, and make a mark on the lumber adjacent to the "24" mark on the ruler.

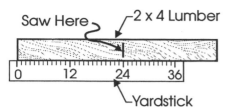

Figure 8-1. Example of a comparison measurement: Use a standardized ruler to find the desired length.

Indirect measurements are made by measuring something other than the actual measurand. Although frequently "second best" from the perspective of accuracy, indirect methods are often used when direct measurements are either difficult or dangerous. For example, you might measure the temperature of a point on the wall of a furnace that is melting metal, knowing that it is related to the interior temperature by a certain factor. Perhaps the most common example of an indirect measurement is human blood pressure. It is measured by measuring the pressure in an occluding cuff placed around the arm. (Research showed that the cuff pressures at two easily detected events are related to the systolic and diastolic arterial pressures.) Direct blood pressure measurement is dangerous because it is an invasive surgical procedure.

Null measurements are made by comparing a calibrated source to an unknown measurand, and then adjusting either one or the other until the difference between them is zero. An electrical potentiometer is such an instrument; it is an adjustable calibrated voltage source. The reference voltage from the potentiometer is applied (Figure 8-2) to one side of a zero-center galvanometer (or one input of a difference-measuring voltmeter), and the unknown is applied to the other side of the galvanometer (or the other input of the differential voltmeter). The output of the potentiometer is adjusted until the meter reads zero difference. The setting of the potentiometer under the nulled condition is the same as the unknown measurand voltage.

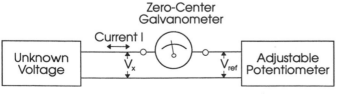

Figure 8-2. Electrical potentiometric comparison measurement. When the unknown voltage V_x equals the reference voltage V_{ref}, the current (I) is zero.

Factors in Making Measurements

How "good" a measurement is involves several concepts that must be understood. Some of the more significant of these are: Error, validity, reliability, repeatability, accuracy, precision, and resolution.

In all measurements there is a certain degree of error present. The word "error" in this context refers to normal random variation, and in no way means "mistakes," "blunders," or SNAFUs. In short order we will discuss error in greater depth. If you make repeated measurements on the same parameter (which is truly unchanging), or if you use different instruments or instrument operators to make successive measurements, you will find that the measurements tend to cluster around a central value (x_0 in Figure 8-3). In most cases, it is assumed that x_0 is the true value, but if there is substantial inherent error in the measurement process, then it may deviate from the true value (x_i) by a certain amount (Δx)—which is the error term. The assumption that the central value of a series of measurements is the true value is only valid when the error term is small. As $\Delta x \to 0$, $x_0 \to x_i$.

The validity of a measurement is a statement of how well the instrument actually measures what it is supposed to measure. A pressure sensor may actually be measuring the deflection of a thin metallic diaphragm (of known area, of course), which is in turn measured by the strain applied to a strain gauge cemented to the diaphragm. What determines the validity of a sensor measurement is the extent to which the measurement of the deflection of that

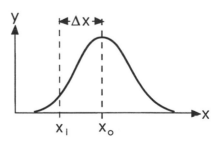

Figure 8-3. The difference between a measured value (x_i) and the actual value (x_0) is Δx, and is a measure of the accuracy of the measurement. We presume that the cumulative distribution of the errors is a normal distribution curve, unless compelling evidence to the contrary exists.

diaphragm relates to applied pressure...and over what range or under what conditions. Many measurement devices exist where the output readings are only meaningful under the right conditions or over a specified range.

The reliability of the measurement is a statement of its consistency when discerning the values of the measurand on different trials, when the measurand may take on very different values. In the case of the pressure sensor discussed above, the deformation of the diaphragm may change its characteristics enough to alter future measurements of the same pressure value. Related to reliability is the idea of repeatability, which means the ability of the instrument to return the same value when repeatedly exposed to the exact same stimulant. Neither reliability nor repeatability are the same as accuracy, for a measurement may be both "reliable" and "repeatable" while being wrong.

The accuracy of a measurement refers to the freedom from error, or the degree of conformance between the measurand and the standard. Precision, on the other hand, refers to the exactness of successive measurements, also sometimes considered the degree of refinement of the measurement. Accuracy and precision are often confused with one another, and these words are often erroneously used interchangeably. One way of stating the situation is to note that a precise measurement has a small standard deviation and variance under repeated trials, while in an accurate measurement the mean value of the normal distribution curve is close to the true value.

Figure 8-4 shows the concepts of accuracy and precision in various measurement situations. In all of these cases, the data form a normal distribution curve when repeatedly performed over a large number of iterations of the measurement. Compare Figures 8-4a and 8-4b. Both of these situations have relatively low accuracy because there is a wide separation between x_o, the measured value, and x_i, the actual value of the measurand. The measurement represented in Figure 8-4a has relatively high precision compared with Figure 8-4b. The difference is seen in the fact that Figure 8-4a has a substantially lower variance and standard deviation around the mean value, x_0; this curve shows poor accuracy but good precision. The variance of Figure 8-4b is greater than the variance of the previous curve, so it has both low accuracy and poor precision.

Somewhat different situations are shown in Figures 8-4c and 8-4d. In both of these distributions, the accuracy is better than the previous cases, because the difference between x_0 and x_i is reduced (of course, a perfect, error-free, measurement has a difference of zero). In Figure 8-4c, the measurement has good accuracy and good precision, while in Figure 8-4d the measurement has good accuracy and but poor precision (its variance is greater than in Figure 8-4c).

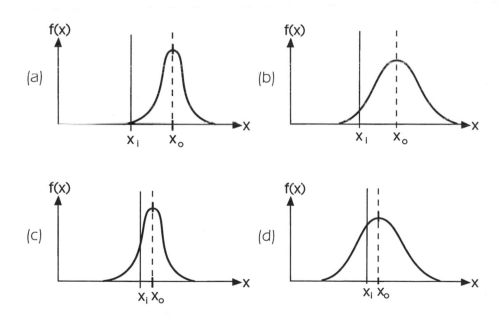

Figure 8-4. (a) High-precision, low-accuracy measurement; (b) low-precision, low-accuracy measurement; (c) good precision and good accuracy measurement; (d) low-precision, high-accuracy measurement.

The standard deviation of the measurement is a good indication of its precision, which also means the inherent error in the measurement. One of my more fondly remembered college professors, Professor George Kohl of The George Washington University, taught us to reduce error in our laboratory experiments by several methods:

1. Make the measurement many times, and then average the results.
2. Make the measurement using different instruments, if feasible.
3. When using instruments such as rulers or meters, try making the successive measurements on different parts of the scale. For example, on rulers the distance between tick marks is not really constant because of manufacturing error. The same is also true of electrical meter scales. Professor Kohl had us measure lengths using different points on the scale as the zero reference point (e.g. on a meter stick use 2, 12, 20, and 30 cm as the zero point), and then average the results. By taking the measurements from different sections of the scale, both individual errors and biases that accumulate will be averaged to a lower overall error.

Resolution refers to the degree to which the measurand can be broken into identifiable adjacent parts. An example can be seen on the standard television test pattern broadcast by some stations in the early morning hours between "broadcast days." Various features on the test pattern will include parallel vertical or horizontal lines of different densities. One patch may be 100 lines per inch, another 200 lines per inch, and so forth up the scale. The resolution of the video system is the maximum density at which it is still possible to see adjacent lines with space between them. For any system, there is a limit above which the lines are blurred into a single entity.

In a digital electronic measuring instrument, the resolution is set by the number of bits used in the data word. Digital instruments use the binary (base-2) number system in which the only two digits are "0" and "1." The binary "word" is a binary number representing a quantity. For example, "0001" represents decimal 1, while "1001" represents decimal 5. An 8-bit binary data word, the standard for many small computers, can take on values from 00000000_2 to 11111111_2, so can break the range into 2^8 (256) distinct values, or 2^8-1 (255) different segments. The resolution of that system depends on the value of the measured parameter change that must occur in order to change the least significant bit in the data word. In digital systems, this resolution is often specified as the "1-LSB" resolution.

Measurement Errors

No measurement is perfect, and measurement apparatus is never ideal, so there will always be some error in all forms of measurement. An error is a deviation between the actual value of a measurand and the indicated value produced by the sensor or instrument used to measure the value. Let me reiterate: Error is inherent, and is NOT the fault of the person making the measurement. Error is not the same as *mistake*! Understanding error can greatly improve our effectiveness in making measurements.

Error can be expressed in either (so-called) "absolute" terms, or using a relative scale. An absolute error would be expressed in terms of "$x \pm x^a$ inches," or some other such unit, while a relative error expression would be "$x \pm 1\%$ inches." Which to use may be a matter of convention, personal choice, or best utility, depending on the situation.

There are four general categories of error: Theoretical error, static error, dynamic error, and instrument insertion error. Let's look at each of these in turn.

Theoretical Error

All measurements are based on some measurement theory that predicts how a value will behave when a certain measurement procedure is applied. The measurement theory is usually based on some theoretical model of the phenomenon being measured, that is, an intellectual construct that tells us something of how that phenomenon works. It is often the case that the theoretical model is valid only over a specified range of the phenomenon. For example, nonlinear phenomenon that have a quadratic, cubic, or exponential function can be treated as a straight-line linear function over small, selected sections of the range.

Alternatively, the actual phenomenon may be terribly complex, or even chaotic, under the right conditions, so the model is therefore simplified for many practical measurements.

An equation used as the basis for a measurement theory may be only a first-order approximation of the actual situation. For example, consider the mean arterial pressure (MAP) that is often measured in clinical medicine and life sciences research situations. The MAP equation used by many people is:

$$\bar{P} = \text{Diastolic} + \frac{\text{Systolic} - \text{Diastolic}}{3}$$

This equation is really only an approximation (and holds true mostly for well people, but not some sick people on whom it is applied) of the equation that expresses the mathematical integral of the blood pressure over a cardiac cycle—that is, the time average of the arterial pressure. The actual expression is written in the notation of calculus, which is beyond most of the people who use the "measurement version" above:

$$\bar{P} = \frac{1}{T} \int P \, dt$$

The approximation works well, but is subject to greater error due to the theoretical simplification of the first equation.

Incidentally, if you work your way through most of undergrad and all of grad school repairing medical instruments, like I did, then you'll understand that it does no good to tell an intensive care unit nurse that the MAP reading on an instrument is correct when it differs from her calculated value. The actual measured MAP is based on the integral calculus equation, while her calculated value is based on the simplification. Because some patients lack a considerable part of the time-dependent pressure curve (which is sometimes why they are in the hospital), the actual MAP will be lower than the approximation value. Sometimes heat, fire, and smoke flow from the tender, loving-care hands of ICU nurses!

Static Errors

Static errors include a number of different subclasses that are all related in that they are always present even in unchanging systems (thus are not dynamic errors). These errors are not functions of the time or frequency variation.

Reading Static Errors. These errors result from misreading the display output of the sensor system. An analog meter uses a pointer to indicate the measured value. If the pointer is read at an angle other than straight on, a parallax reading error occurs. Another reading error is the interpolation error, an error made in guessing the correct value between two calibrated marks on the meter scale (Figure 8-5). Still another reading error occurs if the pointer on a meter scale is too broad, and covers several marks at once.

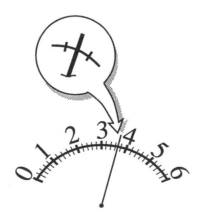

Figure 8-5. Interpolation of measurements is a "guesstimate" of the distance between two "ticks" on the meter face.

A related error seen in digital readouts is the last digit bobble error. On digital displays, the least significant digit on the display will often flip back and forth between two values. For example, a digital voltmeter might read "12.24" and "12.25" alternately, depending on when you looked at it, despite the fact that absolutely no change occurred in the voltage being measured. This phenomenon occurs when the actual voltage is between the two indicated voltages. Error and uncertainty in the system will make a voltage of 12.245 bobble back and forth between the two permissible output states (12.24 and 12.25).

An example where "bobble" is significant is seen in the case of an intensive care unit patient who had an unstable blood pressure. The doctor's orders were to administer a certain medication if the arterial systolic pressure dropped below 90 mmHg. The three-digit numerical display could only read out to the nearest mmHg, so the nurse became confused when the pressure reading bobbled back and forth between 89 and 90; the actual pressure was probably 89.5 mmHg plus or minus a small uncertainty due to measurement error.

Environmental Static Errors. All sensors and instruments operate in an environment, which sometimes affects the output states. Factors such as temperature (perhaps the most common error-producing agent), pressure, electromagnetic fields, and radiation must be considered in some sensor systems.

Characteristic Static Errors. These static errors are still left after reading errors and environmental errors are accounted for. When the environment is well within the allowable limits and is unchanging, when there is no reading error, there will be a residual error remaining that is a function of the measurement instrument or process itself. Errors found under this category include zero-offset error, gain error, processing error, linearity error, hysteresis error, repeatability error, resolution error, and so forth.

Also included in the characteristic error is any design or manufacturing deficiencies that lead to error. Not all of the "ticks" on the ruler are truly 1/16 inch apart at all points along the ruler. While it is hoped that the errors are random, so that the overall average error is small, there is always the possibility of a distinct bias or error trend in any measurement device.

For digital systems you must add to the resolution error a quantization error that emerges from the fact that the output data can only take on certain discrete values. For example, an 8-bit analog-to-digital converter allows 256 different states, so a 0-to-10-volt range is broken into 256 discrete values in 39.06 mV steps. A potential between two of these steps is assigned to one or the other according to the rounding protocol used in the measurement process. An example is the weight sensor that outputs 8.540 volts, on a 10-volt scale, to represent a certain weight. The actual 8-bit digitized value may represent 8.502, 8.541, or 8.580 volts because of the ±0.039 volt quantization error.

Dynamic Errors

Dynamic errors arise when the measurand is changing or in motion during the measurement process. Examples of dynamic errors include the inertia of mechanical indicating devices (such as analog meters) when measuring rapidly changing parameters. There are a number of limitations in electronic instrumentation that fall into this category, as you will find out by consulting any competent book on the subject.

Instrument Insertion Error

A fundamental rule of making engineering and scientific measurements is that *the measurement process should not significantly alter the phenomenon being measured*. Otherwise, the measurand is actually the altered situation, not the original situation that is of true interest. Examples of this error are found in many places. One is the fact that pressure sensors tend to add volume to the system being measured, so slightly reduce the pressure indicated below the actual pressure. Similarly, a flow meter might add length, a different pipe diameter, or turbulence to a system being measured. A voltmeter with a low impedance of its own could alter resistance ratios in an electrical circuit and produce a false reading. These errors can usually be minimized by good design. No measurement device has zero effect on the system being measured, but you can reduce the error to a very small value by appropriate selection of methods and devices.

Dealing With Measurement Errors

Measurement error can be minimized through several methods, some of which are lumped together under the heading "procedure" and others under "statistics."

Under "procedure" you can find methods that will reduce, or even minimize, error contributions to the final result. For example, in an electrical circuit, use a voltmeter that has an extremely high input impedance compared with circuit resistances. The idea is to use an instrument (whether a voltmeter, a pressure meter or whatever) that least disturbs the thing being measured.

A way to reduce total error is to use several different instruments to measure the same parameter. In Figure 8-6 we see an example where the flow in a hydraulic circuit is being measured by three different flowmeters: S1, S2, and S3. Each of these instruments will produce a result that contains an error term decorrelated from the error of the others and not biased (unless, by selecting three identical model meters we inherit the characteristic error of that type of instrument). We can estimate the correct value of the flow rate by taking the average of the three:

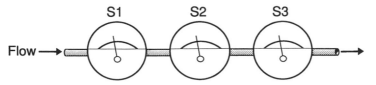

Figure 8-6. It is sometimes possible to improve the measurement by using three independent meters to measure the same thing, in this case a fluid flow in a pipe.

$$S_0 = \frac{S_1 + S_2 + S_3}{3}$$

You must be careful to either randomize the system in cases where the sensor or instruments used tend to have large error terms biased in one direction, or calibrate the average error so that it may be subtracted out of the final result.

Error Contributions Analysis

An error analysis should be performed in order to identify and quantify all contributing sources of error in the system. A determination is then made regarding the randomness of those errors, and a worst-case analysis is made. Under the worst case, you assume that all of the component errors are biased in a single direction and are maximized. You then attempt to determine the consequences (to your purpose for making the measurement) if these errors line up in that manner, even if such an alignment is improbable. The worst-case analysis should be done on both the positive and negative side of the nominal value. An error budget is then created to allocate an allowable error to each individual component of the measurement system in order to ensure that the overall error is not too high for the intended use of the system.

If errors are independent of each other, and are random rather than biased, and if they are of the same order of magnitude, then you can find the root of the sum of the squares (rss) value of the errors, and use it as a composite error term in planning a measurement system. The rss error is:

$$\varepsilon_{rss} = \sqrt{\Sigma \varepsilon_i^2}$$

The rss error term is a reasonable estimate or approximation of the combined effects of the individual error components.

A collection of repetitive measurements of a phenomenon can be considered a sampled population, and treated as such.

If we take N measurements (M_1 through M_n) of the same value, and average them we get:

$$\overline{M} = \frac{M_1 + M_2 + M_3 + \ldots + M_n}{N}$$

This value can be reported as "the" value for the measurement, but does not correctly address the issue of error. For this purpose we add a "fudge factor" by quoting the standard error of the mean, or

$$\sigma_{\overline{m}} = \frac{\sigma_m}{\sqrt{N}}$$

which should be reported in the result as:

$$M = \overline{M} \pm \sigma_{\overline{m}}$$

Any measurement contains error, and this procedure allows us to estimate that error, and thereby understand the limitations of that particular measurement.

Some measurement procedures suggest themselves immediately from the nature of the phenomenon being measured. In other cases, however, there is a degree of ambiguity in the process, and it must be overcome. Sometimes the ambiguity results from the fact that there are many different ways to define the phenomenon, or perhaps no way is established. In cases such as these, you might wish to resort to an *operational definition*, a procedure that will produce consistent results from measurement to measurement, or when measurements are taken by different experimenters.

An operational definition, therefore, is a procedure that must be followed, and specifies as many factors as are needed to control the measurement so that changes can be properly attributed only to the unknown variable—for example, the addition of salt to the water in the example above. The need for operational definitions (as opposed to absolute definitions) arises from the fact that things are only rarely so neat and clean as to suggest their own natural definition; ambiguity reigns. By its very definition, the operational definition does not ask "true" or "false" questions, but rather it asks "what happens under given sets of assumptions or conditions." What an operational definition can do for you is to standardize a measurement in a clear and precise way so that it remains consistent across numerous trials. Operational definitions are used extensively in science and technology. When widely accepted, or put forth by a recognized authority, they are called *standards*.

An operational definition should embrace what is measurable quantitatively, or at least in nonsubjective terms. For example, in measuring the "saltiness" of saline solution (salt water), you might taste it and render a subjective judgment such as "weak" or "strong." A better way would be to establish an operational definition that calls for you to measure the electrical resistance of the saline under certain specified conditions:

1. Immerse two 1-cm diameter circular nickel electrodes, spaced 5 cm apart and facing each other, to a depth of 3 cm into a 500-ml beaker of the test solution.

2. Bring the solution to a temperature of 4°C.

3. Measure the electrical resistance (R) between the electrodes using a *Snotz* Model 1120 digital ohmmeter.

4. Find the conductance (G) by taking the reciprocal of resistance ($G = 1/R$).

You may come up with a better definition of the conductance of saline solution that works for some peculiar situation. Keep in mind that only rarely does a preferred definition suggest itself naturally.

The use of operational definitions results in both strengths and weaknesses. One weakness is that the definition might not be honed fine enough for the purpose at hand. Sociologists and psychologists often face this problem because of difficulties in dealing with nonlinearities such as human emotions. But such problems also point to a strength. We must recognize that scientific truth is always tentative, so we must deal with uncertainties in experimentation. Sometimes, the band of uncertainty around a point of truth can be reduced by using several operational definitions in different tests of the same phenomenon. By taking different looks from different angles, we may get a more refined idea of what's actually happening.

When an operational definition becomes widely accepted, and is used throughout an industry, it may become part of a formal standard or test procedure. You may, for example, see a procedure listed as "performed in accordance with NIST XXXX.XXX" or "ANSI Standard XXX." (NIST stands for National Institute for Standards and Technology, formerly called the National Bureau of Standards. ANSI is American National Standards Institute.) These notations mean that whoever made the measurement followed one or another of a published standard.

References

N.R. Campbell, *Foundations of Science*, Dover (New York, 1957). Cited in John Mandel, *The Statistical Analysis of Experimental Data*, John Wiley & Sons (New York, 1964). Dover paperback edition 1984.

E.E. Herceg, *Handbook of Measurement and Control*, Schaevitz Engineering, (1972, Pennsauken, NJ).

Probability—A Scientific "Game of Chance"?

PROBABILITY DEALS in likelihoods rather than absolutes. It allows us to anticipate how often to expect certain outcomes in situations where more than one outcome is possible, for example, in the flipping of a coin or the rolling of dice (the classical examples). Probability theory is also used in science, engineering, and measurements.

It is amusing to note that, while probability theory is used very effectively today in science, medicine, engineering and a host of other modern technical disciplines, it was in predicting games of chance—gambling—that probability theory got its start. It seems that many of the better-known mathematicians of the sixteenth and seventeenth centuries did some of their best theoretical work in the service of rich clients who had a penchant for gaming. Even today one occasionally hears a story of a bright math wizard walking away from the casino or race track with a small fortune. (But don't bet on it happening too often—casinos employ statisticians, too!)

Probability problems use the letter P to denote "probability of...," with an argument following the P in parenthesis or brackets to indicate what events are being discussed. For example, "$P(x)$" means the "probability of event x occurring." In flipping coins, you might write "$P(heads)$" to denote the probability of a heads occurring on a flip, or "P(tails)" to mean the opposite case.

The value of P is a fraction (sometimes written in decimal form) between 0 and 1, where $P(x) = 0$ means that there is no possibility whatsoever that event x will occur, and $P(x) = 1.00$ means there is no possibility whatsoever that x will not occur. In other words, a probability of zero means no occurrence, while a probability of 1.00 indicates a certainty that x will occur. Values of P between 0 and 1 are an indication of the relative likelihood of x occurring. For example, if the likelihood of x occurring is 0.5, that means that half the time you can expect x to occur, and half the time x will not occur. It does not predict the absolute outcome of any specific future event, only the likelihood of each possible outcome.

Suppose we know that 4.5% of the students in a school wear dental braces. This means that 4.5 students out of every 100 are wearing the braces, which is a proportion of 0.045. What is the probability of randomly selecting a student with braces from the entire school student body? The probability is P(braces) = 0.045, or to express it differently take the reciprocal: $1/0.045 = 22.22$. That means that there is a 1 in about 22 chance of randomly selecting a wearer of braces out of the entire student body.

The probability 0.045 only holds true if: (a) all students have an equal chance of being selected (in other words, the selection process is truly random), and (b) the entire student body was in the pool to be chosen.

Different aspects of probability theory are used to cover two broad classes of events: simple and compound. Simple events cannot be reduced further into more possible events; for example, the flip of a coin can only result in heads (H) or tails (T), or a rolled die can only result in 1, 2, 3, 4, 5, or 6. You cannot take the "5" face of a die and dissect it into smaller elements of "fiveness". . . it always comes up just "5" or not at all.

If a simple event can have more than one outcome, and if all possible outcomes are equally likely, then the probability of any one of them occurring in a single fair trial is the proportion that outcome has of all possible outcomes. If there are N different possibilities of event A, and n outcomes are "successes," then the probability of event A, that is $P(A)$, is n/N. The probability of event A is a fraction between 0 and 1: i.e., $0 < P(A) \leq 1.00$. Thus, for the flipping of a coin, there are two possibilities (H or T), so the probability of "calling it" on any one flip is 1 "success" over 2 possibilities, or $P(A) = \frac{1}{2} = 0.50$. This fact is sometimes called the *First Law of Probability*.

Compound events can be decomposed into two or more possible events; for example, the probability that rolling a die will result in either an even number (2, 4, or 6) or odd number (1, 3, or 5). The probability of the event is the sum of the probabilities of the individual possibilities. For example, the possibility of rolling a 2, 4, or 6 (i.e., rolling an even number):

$$P(\text{even}) = P(2) + P(4) + P(6)$$

$$P(\text{even}) = \frac{1}{6} + \frac{1}{6} + \frac{1}{6} = \frac{3}{6} = \frac{1}{2}$$

Constructing a Probability Diagram

For the simple case of a coin flip, it is relatively easy to see the probability situation for multiple trials by using a graphical device called a probability diagram (See Figure 9-1). On the first flip of the coin, only one of two possible outcomes (H or T) is possible, and each has a probability of 0.5. If the result is heads (H), then the graph takes us into results Group A; but if it is tails (T) we are results Group B. On the second flip there are still two possibilities (H and T), so the "tree" branches again...and the process repeats over and over again.

We can use Figure 9-1 to write all of the possible combinations for as many flips as we have room to graph. In the case of Figure 9-1, there are three flips, so (starting at the first flip and counting down each possible path) the possibilities are: HHH (path shown by arrow in Figure 9-1), HHT, HTH, HTT, THH, THT, TTH, and TTT.

Figure 9-1. Probability tree for the coin flip problem.

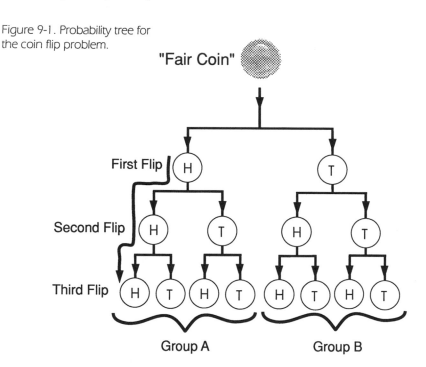

Methods of Classifying Probability Events

Other methods for classifying probability events are: (1) Equally Likely, (2) Long Term Averages, and (3) Baysian. The Equally Likely probability is applied to individual events such as the flipping of a coin. According to LaPlace's Principle, we must consider all possibilities of the event to be equally likely unless there is compelling evidence to indicate otherwise (for example, an exceedingly lucky run of rolled dice, all the same number, might lead one to suspect that the dice are loaded—

although extremely low-probability long runs actually do occur). The single event will have an equally likely chance of producing an outcome of any of the possibilities.

The Long Term Averages concept is merely a statement of the history of repeated trials of an event. For example, if you flip a fair coin 1000 times, you will find a nearly equal number of heads and tails. Note the term "nearly," for it is not strictly necessary that the actual trial wind up with a demonstrated historical probability of exactly 0.50 for the single event probability to be 0.50.

The Baysian class is a way of dealing with ambiguous cases. In some situations we know the probability of an event, A, occurring given that another event, B, has also occurred. For example, we might know the probability of a person having a certain disease if they also had a positive test for that disease. These situations are denoted "$P(A \mid B)$," meaning "the probability of A given that B has occurred." Baysian probabilities allow us to ask the alternate question, $P(B \mid A)$, or "What is the probability of B being true if A is also true." A cancer test, for example, may yield true positives on 85% of cancer victims, but also gives false negatives 15% of the time. We may ask "What is the probability of cancer given a positive test?"

Second Law of Probabilities— Averages of Large Trials

If an event can have more than one outcome, and if all possible outcomes are equally likely, then the results will nearly always vary somewhat from the calculated ideal probability. But for a large number of trials, the variation will be smaller if all of the events are truly equally likely. A European scientist in the nineteenth century rolled a single pair of dice, one red and one white, more than 20,000 times over a period of 43 years. He kept detailed records of all trials. In the end, it was noted that the white die obeyed the law of large numbers, while on the red die the "6" face came up 38% more often than the "4" face. What did this extensive investigation prove? Nothing, except that the red die was a bit lopsided. (This story is told in Gary Smith's book *Statistical Reasoning*—see the references.)

For the law of averages in large numbers to be valid, all possibilities must be equally likely. That is, they all must have an equal chance of occurrence in a fair test. In coin flipping, a "fair coin" is one that is not bent and is in no way weighted towards either heads or tails. In dice, the "fair die" is one that is not loaded. Gamblers of old were known to insert a shotgun pellet into the die, placed off center in such a way to yield a preferred direction. In those "loaded" dice, the faces away from the piece of lead shot would come up disproportionately more often than other faces. Irregular manufacture also results in lopsided results, as in the case of the die used by the European scientist who recorded a lifetime of dice rolls.

In playing cards, a "fair deck" is one in which all of the cards are of uniform dimensions, including thickness, are unbent, unmarked, and are well shuffled at least five times. The order of a new deck of cards is anything but random, but after proper shuffling the order is randomized.

It is a common fallacy to assume that the Law of Large Numbers of Trials requires the proportion of each possibility to be exact. For example, above we used the coin flip case as an example. The probability of seeing $P(H)$ or $P(T)$ be exactly 0.5000 after, say, 10,000 trials is extremely low, but the probability that the actual value is *near* 0.5 is extremely high. Gamblers who think that a run of heads or tails must be exactly counterbalanced later on are betting a fool's bet...as are dice shooters who fail to realize that on a large number of throws the proportion of any one number on the die will be close to, but unlikely to be exactly, 1/6. Gamblers fall all over each other to place fool's bets because of ignorance of probability theory.

Third Law of Probability—Law of Addition

If something can have more than one outcome, and if all possible outcomes are equally likely, then the probability of alternative results on a *single* trial will be the sum of their individual probabilities. We saw this law in operation in the discussion of compound events above, but let's go over it one more time.

Earlier we considered the probability of rolling a fair die and coming up with an even number. It is known that the possibilities include six different "faces" of the die—1, 2, 3, 4, 5, or 6—of which three—2, 4, and 6—are even numbers. Because there are six possible outcomes, the probability of any one of them is 1/6, or about 0.1667. For the case where all we care about is the probability of an even number, $P(\text{even}) = P(2,4, \text{ or } 6)$, the probabilities are:

$$P(\text{even}) = \frac{1}{6} + \frac{1}{6} + \frac{1}{6}$$

or,

$$P(\text{even}) = 0.1667 + 0.1667 + 0.1667 = 0.5001 = 0.5$$

In other words, the probability of rolling an even number is very close to 0.5. Notice also that the proportion of even numbers to all possible numbers is, not surprisingly, 0.5. Because there are also three possible odd numbers, the value of $P(\text{odd})$ is also 0.5.

The probability of any such event occurring is not dependent on past trials. In other words, a coin does not have some kind of computer memory to let it know how many "heads" or "tails" came up in the past trials. Each trial, or future number of trials, has an independent probability that in no way depends on the *past* history of the trials. History does not predict the next flip of a coin or roll of the dice!

A deck of 52 playing cards has four aces: ♥, ♦, ♣, and ♠. Assuming that the deck is well shuffled, what is the probability of drawing *any* ace? Because there are four possibilities out of 52, the probability is:

$$P(\text{ace}) \ = \ \frac{4}{52} \ \approx \ 0.0769$$

What is the probability of drawing an ace on the next trial? It is not the same 4/52 because now there are only 51 cards in the deck. If no ace was drawn on the first trial, then there are still 4 aces remaining in the deck, so the number is 4/51, or 0.07843. If, on the other hand, an ace was drawn on the first trial, then the probability of drawing another is 3/51 (0.0588) because now there are only 3 aces in the deck of 51. Shortly we will look at the probability of drawing all 4 aces in a row.

A similar problem is the bead experiment. Suppose we have a large barrel full of beads. There are 1,000 red beads, and 4,000 white beads, for a total of 5,000 beads. The beads are well mixed, the barrel having been rotated a few dozen times, and then shaken on a paint-shaking table. The probability of reaching blindly into the barrel and pulling out a white bead is 4000/5000, or 4/5 (i.e., 0.8). Alternatively, the probability of obtaining a red bead is 1000/5000, or 1/5 (i.e., 0.2). Note that the combination of $P(\text{red})$ and $P(\text{white})$ is 1.00. Because only red and white beads are in the barrel, there is a probability of 1.00 that either a red or a white will be drawn...regardless of whether the person pulling the beads is blindfolded or not.

Fourth Law of Probabilities— Law of Multiplication

Whenever something can have more than one outcome, if all possible outcomes are equally likely, then the probability of any particular *combination* of outcomes on two or more independent trials is the product of their respective probabilities.

If each event is truly independent, and the order is not specified, then the probability of *either* event occurring in a single trial is the sum of their probabilities. This situation is controlled by the addition law, stated above. What is the probability of getting a particular combination on two or more successive trials? For example, what is the probability of rolling a 3 and a 4 (order not important) on two throws of the die? For this probability we turn to the multiplication law:

$$P(A \text{ and } B) = P(A) \times P(B)$$

or,

$$P(3 \text{ and } 4) = P(3) \times P(4)$$

$$P(3 \text{ and } 4) = \frac{1}{6} \times \frac{1}{6} = \frac{1}{36}$$

But that's not the whole story. There are two different ways that we can roll both a 3 and a 4 when order is not specified:

WAY	1st Throw	2nd Throw
A	3	4
B	4	3

Probability for Way A (Multiplication law):

$$P(3) \times P(4) = ?$$

$$\frac{1}{6} \times \frac{1}{6} = \frac{1}{36}$$

Probability for Way B (Multiplication law):

$$P(3) \times P(4) = ?$$

$$\frac{1}{6} \times \frac{1}{6} = \frac{1}{36}$$

Probability of either Way A or Way B on a single trial (Addition law):

$$P(A \text{ and } B) = P(A) + P(B)$$

$$\frac{1}{36} + \frac{1}{36} = \frac{1}{18}$$

Now consider a standard 52-card deck of ordinary playing cards. There are four different types of ace: ♥, ♦, ♣, and ♠. What is the probability of drawing an ace (any ace) from a well-shuffled deck on a single try? There are four possibilities out of 52 cards, so the probability is 4/52, or 0.0769 (1 in 13). The probability of drawing a specific ace (e.g., the ace of spades, ♠) is 1/52, or 0.019 (1 in 52). The probability of drawing all four aces in four successive trials is:

$$P(\text{All Four Aces}) = \frac{4}{52} \times \frac{3}{51} \times \frac{2}{50} \times \frac{1}{49}$$

$$P(\text{All Four Aces}) = 0.769 \times 0.0588 \times 0.06 \times 0.020$$

$$P(\text{All Four Aces}) = 0.00000543$$

or, about 1 chance in more than 184,000 (that is, $1/P$). On the first draw, all four aces are still in the deck of 52 cards, so n/N is 4/52. On the second draw, one ace is missing and the deck is reduced to 51 cards, so n/N is 3/51, and so forth. This case is an example of *sampling without replacement* (see Chapter 12 for more information on sampling).

Gambler's Self-Delusion

A pair of obstetrical nurses in the Labor and Delivery department of the local hospital noted that the last 21 babies born in the unit were all girls. There are always two possible outcomes in a birth: Either it is a boy (B) or a girl (G). Thus, the individual birth probability is $P(B) = \frac{1}{2}$ and $P(G) = \frac{1}{2}$, and, the probability of 22 in a row being of one sex (all boys or all girls) is (by the multiplication law):

$$P(\text{All Girls}) = P(G_1) \times P(G_2) \times P(G_3) \dots P(G_{22})$$

$$P(\text{All Girls}) = \frac{1}{2} \times \frac{1}{2} \times \dots \times \frac{1}{2}$$

$$P(\text{All Girls}) = \left(\frac{1}{2}\right)^{22}$$

$$P(\text{All Girls}) = 2.38 \times 10^{-7}$$

Which is a proportion of 0.00000238, or about 1 chance in 4,194,304. Sounds like a great gambling odds, right? Wrong! If the nurses bet their life savings on the next one being a boy— ending the run—they could very easily lose. Why? Because each delivery is an independent event with two possible out-comes (G or B), so the probability of the next birth being a girl (or a boy for that matter) is 0.50. If you bet on the run of all 22, then you can consider it the "event," and the odds are indeed one in more than 4,000,000.

Probability in Compound Events— Addition Law

The simple addition law that we discussed previously works when all outcomes are equally likely. In some compound event situations, however, the outcomes are often not equally likely, and for these we can use a related but slightly different approach. To find the probability of either event A or event B:

$$P(A \text{ or } B) = P(A) + P(B) - P(A \text{ and } B)$$

Let's return to our deck of cards. There are 13 hearts, 13 clubs, 13 spades, and 13 diamonds. There are also 12 face cards—King, Queen, and Jack—with one in each of the four suits. What is the probability of drawing a single card in a single trial that is both a heart and a face card? Try this:

$$P(\text{heart or face}) = P(\text{heart}) + P(\text{face}) - P(\text{heart and face})$$

There are 13 hearts in a deck of 52, so $P(\text{heart}) = 13/52$. Similarly, there are 12 face cards, so $P(\text{face}) = 12/52$, of which 3 are also hearts, $P(\text{heart and face}) = 3/52$. Therefore:

$$P(\text{heart or face}) = \frac{13}{52} + \frac{12}{52} - \frac{3}{52}$$

$$P(\text{heart or face}) = 0.25 + 0.231 - 0.058 = 0.423$$

Or consider another situation. In a research and development department of a certain company there are 200 engineers, of which 80 of them are females. Among the engineers, 12 of the males and 16 of the females hold Ph.D degrees. What is the probability of randomly selecting a single engineer who is either female or a Ph.D?

$$P(\text{female or Phd}) = P(\text{Female}) + P(\text{Phd}) - P(\text{Female and Phd})$$

The probability of an engineer in this company being female is 80/200, while the probability of an engineer being a Ph.D is $(12 + 16)/200$, or 28/200. The probability of being both female and Ph.D is 16/200.

$$P(\text{Female or Phd}) = \frac{80}{200} + \frac{28}{200} - \frac{16}{200}$$

$$P(\text{Female or Phd}) = 0.4 + 0.14 - 0.08 = 0.46$$

We can conclude that events A and B are independent events and mutually exclusive (i.e., if one occurs then the other won't) if it is found that

$$P(A \text{ and } B) = 0$$

Conditional Probabilities

The probability that A occurs given that B also occurs is called a conditional probability; the denotation is $P(A \mid B)$.
The probability of $P(A \mid B)$ is:

$$P(A \mid B) = \frac{P(A \text{ and } B)}{P(B)}$$

Perhaps an example will show what this means.

Cancer Detection Problems

A cancer test is known to be 95% reliable. In other words, 95% of all positive tests are from patients with real cancer ("true positives"), while 5% of positives are from patients with no cancer ("false positives"). Further, 95% of all negatives are from patients who have no cancer ("true negatives"), while 5% of the negatives are from patients who actually did have cancers ("false negatives"). Suppose that there is a population in which 0.5 percent of all people have cancer. This is a proportion of 0.005. By the multiplication law:

$$P(+\text{test and cancer}) = P(+\text{test}) \times P(\text{cancer})$$

$$P(+\text{test and cancer}) = 0.95 \times 0.005 = 0.00475$$

But we have to allow for false positives, which can occur in 5% of the tests. The "universe" for false positives reduces to $(100\%) - (0.5\%) = 99.5\%$, or a proportion of 0.995. For a false positive rate of 5%, the proportion is 0.05, so:

$$0.995 \times 0.05 = 0.04975$$

The probability that any one person, with or without cancer, will have a positive test $[P(+\text{test})]$ is the sum of these:

$$0.00475 \times 0.04975 = 0.05450$$

The probability that a positive test is a true positive test is the proportion:

$$\frac{P(\text{cancer}) + P(+\text{test})}{P(+\text{test})} = \frac{0.00475}{0.05450} = 0.09$$

The Case of the Spurious(?) Coffee Expert

A veteran coffee drinker claims that he can tell the difference between cups of coffee poured before sugar is added (Way A), and cups made with the sugar placed in the bottom of the cup before the coffee is added (Way B). A test is scheduled to prove or disprove this claim. A series of ten cups are made, five each for Ways A and B. The probability of getting either A or B purely by chance on the first trial is 5 out of 10, or 5/10 (i.e., 0.5). For the entire trial, guessing right purely by chance produces a probability of:

$$P(\text{Right}) \;=\; \frac{5}{10} \;\times\; \frac{4}{9} \;\times\; \frac{3}{8} \;\times\; \frac{2}{7} \;\times\; \frac{1}{6}$$

$$0.5 \times 0.444 \times 0.375 \times 0.286 \times 0.167 = 0.004$$

Thus, the probability of correctly guessing, rather than getting it right from pure skill, is 0.004, or about 1 in 250. If the expert does considerably better than this, then we might want to accept the claim.

Some facts to ponder: (1) A and B are independent when $P(A \mid B) = P(A)$, and (2) $P(A \mid B) = P(A)$, then $P(B \mid A) = P(B)$.

Subtraction Rule

The probability of an event occurring is equal to 1 less the probability that it will NOT occur; i.e., $P(A) = 1 - P(\text{NOT-}A)$. The subtraction rule is especially useful in cases where the existence of an event is either very difficult or impossible to determine. If there are only two possibilities, A and NOT-A, then $P(A) + P(\text{NOT-}A) = 1.00$, so we can use the NOT-A case to predict $P(A)$.

Consider a machine consisting of five parts: A, B, C, D, and E. It is known from experience and design considerations that these parts are very reliable, and have a small failure rate.

Table 9-1 below shows the proportion of good parts to failing parts:

TABLE 9-1	PART	PERCENTAGE WORKING
	A	0.999
	B	0.98
	C	0.998
	D	0.97
	E	0.98

For each of these we know the probability of using a good part to assemble the final machine. For example, P(A Good) = 0.999. The multiplication rule gives us the probability of randomly selecting all five parts good:

$$P(\text{All Good}) = P(\text{A Good}) \times P(\text{B Good}) \times P(\text{C Good}) \times$$

$$P(\text{D Good}) \times P(\text{E Good})$$

or,

$$P(\text{All Good}) = (0.999)\,(0.98)\,(0.998)\,(0.97)\,(0.98) = 0.9288$$

The probability of assembling a nonworking machine due to bad parts is found from the subtraction rule:

$$P(\text{Nonworking}) = 1 - P(\text{All Good})$$

$$P(\text{Nonworking}) = 1 - 0.9288 = 0.0712$$

References

Gary Smith, *Statistical Reasoning*, Allyn & Bacon (Needham Heights, MA, 1991).

10

How The Cookie Crumbles— Data Distribution

ROPERLY interpreting scientific data requires us to understand a few things about data distribution. Much of the real knowledge gleaned from scientific experiments is due to proper interpretation of numerical data. The issues raised by data distribution become particularly acute when measuring things, because of the random variation that always creeps into the process. If the same fluid flow rate is measured by the same instrument several times in a row, the odds are pretty good that there will be some difference in the numbers recorded. Similarly, if different instruments are used to make the same measurement, the results are likely to be different.

There is also a tremendous amount of variation in natural phenomena themselves. Just look around you in any large crowd to note how many different heights there are among the people. In one of my engineering classes, there was one man who was only 4 ft 6 in. high, and another (a basketball player) who was 6 ft 9 in. tall.

When you examine any natural phenomenon, measurement error aside, there will be a tremendous amount of variability in the numbers. But when you plot all the data on a graph, so that the relative frequency of occurrence of each event can be seen, you will see that there is a certain pattern to the values.

In the case of my fellow engineering students, had anyone plotted the heights of the students on a relative frequency graph, it would have been apparent that those 4 ft 6 in. and 6 ft 9 in. guys were not close to the average height, which was a bit under 6 ft. That graphical pattern of data points discloses the *distribution* of the data.

Discrete and Continuous Data Distributions

Discrete data distributions are the distributions of counted things; that is, the numbers of males and females in litters of pups, the faces of dice that come up when the dice are rolled, the number of peas in a pod, and so forth. When you roll a single die, you only see one of the six faces on any one trial (1, 2, 3, 4, 5, or 6). Unless you are rolling some kind of weird quantum die, only these numbers turn up. You will never see 0.1, 4.5 from any number of dice, or 8 from any single die.

Figure 10-1a shows the distribution of the probabilities of each face of the die coming up. It consists of a series of six spikes, each with a probability of 1/6, located at each of the six counting numbers that represent the various faces of the die. It is not proper to represent this type of distribution as anything other than a series of spikes. Connecting the spikes does not show any information at all. Indeed, connecting the tops of the spikes would imply that a value such as "4.34232" is possible, which it clearly is not.

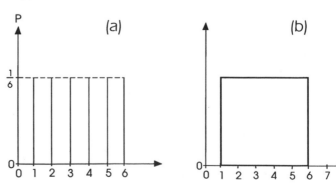

Figure 10-1. (a) Probability chart for throwing a single die (discrete distribution); (b) continuous distribution.

If a data point can actually take on any value between range limits, then its distribution is a continuous curve, and is called (not surprisingly) a *continuous data distribution*. Figure 10-1b shows a continuous distribution with a range the same as that for the dice in Figure 10-1a. The independent variable can take on any value within the domain 1 to 6, and (what this graph tells us) all possible values have an equally likely chance of occurring.

Although the principal data distributions can be classified as either discrete or continuous, you will sometimes see other types. For example, in computer-acquired data, there are situations in which the independent variable is continuous while the dependent variable is discrete. Similarly, there are situations in which both the independent and dependent variables are discrete. If you are into electronics, then you might recognize that continuous functions are "analog signals" and discrete functions are "digital signals." The case where a continuous independent variable is coupled with a discrete dependent variable is called *digitized analog data*; and the case where both variables are digitized is *sampled and digitized analog data*. But more of those classes later, in another chapter.

The frequency of occurrence data for a single die showed a series of equal height spikes at each counting number from 1 to 6. When two dice are thrown together the possible values range from 2 to 12, although not every number in the range has an equal probability of occurring. This change in probability of occurrence comes from the fact that some members of the middle portion of the range can be formed from more than one combination of dice. The number "7," for example, can be formed from 3+4, 5+2, or 6+1, so it has a higher probability of occurring than either 2 or 12.

The distribution for a large number of trials of two dice are shown in Figure 10-2a. Notice that the distribution is triangular, and is centered on 7. If a larger number of dice are thrown the central value changes, but the curve begins to smooth out, as shown in Figures 10-2b and 10-2c, and forms the familiar "bell-shaped curve." When the bell-shaped curve is continuous, the area under the curve represents the probability of occurrence. Several different types of data distribution form into the general bell shape, but the most common is the *normal distribution curve*.

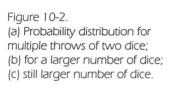

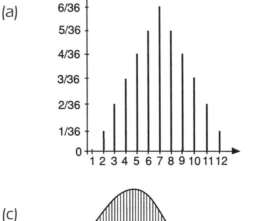

(a)

Figure 10-2.
(a) Probability distribution for multiple throws of two dice;
(b) for a larger number of dice;
(c) still larger number of dice.

(b)

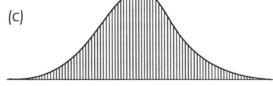

(c)

The Normal Distribution Curve

The English scientist Sir Francis Galton (1822–1911) apparently was crazy about the normal distribution curve (Figure 10-3), exclaiming that the Greeks would have deified it, had they known of it. In the generation before Galton, French scientist Pierre-Simon Laplace (1749–1827) became enamored of the curve and spent a lifetime studying it. Today in France, the normal distribution curve is commonly called the *Laplacian* because of his efforts. The German mathematician Karl Frederich Gauss (1777–1855) applied the curve to so many different phenomenon that it is commonly called the *Gaussian distribution* everywhere else but France, even though it was discovered by another mathematician.

The origins of the normal distribution curve are found in London, in 1733, in the work of a French refugee named Abraham de Moivre (1667–1754). Moivre was working for gentlemen gamblers (what else!) when he discovered the curve from studies of probability theory. He discov-ered that this curve held true for

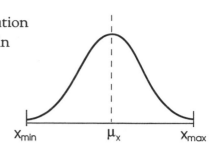

Figure 10-3. Normal distribution curve.

many gaming activities. According to de Moivre, for binomi-ally distributed data—i.e., when $P = 0.50$ for each trial—the curve is a very close approximation of the entire data set for large numbers of trials.

In Figure 10-3 the central peak (μ_x) represents the mean average, while x_{min} and x_{max} represent the minimum and maxi-mum values, respectively. The mean value tells us something about the data, but it is an incomplete view of the data set in question. Likewise, the minimum and maximum values tell us something about the data. The *spread* of the data, $x_{max} - x_{min}$, gives us the range of possible values, but still the view of the data is not sufficient to characterize it. For a better under-standing of the data represented by the normal distribution curve we turn to the important concepts of deviation, variance and standard deviation.

The deviation of the data is defined as the distance of any given data point x_i from the mean value, μ_x, or:

$$\text{Deviation } (x) = x_i - \mu_x$$

If we sum all of the deviations together and take the aver-age, then we find that the average deviation is always zero. In other words, a deviation between μ and $+x_i$ is balanced by a deviation between μ and $-x_i$.

The deviation is important, but does not satisfy the require-ment for insight into the behavior of the data set at large. It only deals with the properties of any given point in the range.

The variance of the data, a measure of its dispersion, is the average squared deviation from the mean, or

$$\text{Variance} = \frac{\sum_{i=1}^{n}(x_i - \mu_x)^2}{n}$$

The variance of a data population is usually denoted by the lower case Greek letter *sigma* squared (σ^2).

EXAMPLE

A jeweler is making "casting nuggets" from 0.925-fine sterling silver round wire. He wants each to be 850 milligrams in mass, but is using *length* to determine the weight. If the wire is uniform, and the method used to measure length accurate, then this method should work properly. After five nuggets are made the jeweler determines the average and variance. The values are: $x_1 = 871$; $x_2 = 872$; $x_3 = 866$; $x_4 = 871$; $x_5 = 869$; $n = 5$.

Solution:

1. Calculate the mean:

$$\bar{x} = \frac{871 + 872 + 866 + 871 + 869}{5}$$

$$\bar{x} = \frac{4349 \text{ mg}}{5} = 869.8 \text{ mg}$$

2. Calculate the variance, σ^2:

$$\sigma^2 = \left(\frac{1}{5}\right)[(871 - 869.8)^2 + (872 - 869.8)^2 +$$

$$(866 - 869.8)^2 + (871 - 869.8)^2 + (869 - 869.8)^2]$$

$$\sigma^2 = \left(\frac{1}{5}\right)[(1.2)^2 + (2.2)^2 + (-3.8)^2 + (1.2)^2 + (-0.8)^2]$$

$$\sigma^2 = \frac{1}{5}[1.44 + 4.84 + 14.44 + 1.44 + 0.64] = \frac{22.8}{5} = 4.56$$

The mean of the nuggets was 869.8 mg with a variance of ±4.56 mg.

If sampled data is used in the calculation of variance, rather than the entire population, it is customary to divide the numerator by $n-1$ rather than n, and use s^2 to denote variance and $\bar{x}$ rather than μ for the mean:

$$s^2 = \frac{\sum_{i=1}^{n} (x_i - \bar{x})^2}{n-1}$$

(See Chapter 12 for more details on sampled data.)

A problem with variance is that it can climb to very high values with even moderate dispersion, so we often prefer to use instead the *standard deviation* (σ), which is the square root of variance. The equations become:

For the population case:

$$\sigma = \sqrt{\frac{\sum_{i=1}^{n} (x_i - \bar{x})^2}{n}}$$

For sampled data:

$$s = \sqrt{\frac{\sum_{i=1}^{n} (x_i - \bar{x})^2}{n-1}}$$

In the preceding case, the variance σ^2 was 4.56 mg, so the standard deviation is:

$$\sigma = \sqrt{\sigma^2} = \sqrt{4.56} = 2.14$$

The standard deviation is a measure of the spread—or dispersion—of the data, as you can see graphically in Figure 10-4. Curve number 1 has a large variance and standard deviation, so it tends to spread out over a larger area. Curve number 2 has a smaller standard deviation, so it narrows compared with curve number 1. Curve number 3 has a small standard deviation so is sharper than the other two curves. Of these, curve number 3 is the most precise, and would make the best estimate of the mean value. The standard deviation tends to drop as the number of data points gets large. In sampling, the ideal goal is to reduce the standard deviation to the smallest value possible. Similarly, when making measurements you need to reduce variability due to the measurement process itself to as small a value as possible.

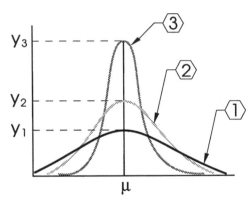

Figure 10-4. Three normal distributions for (1) large data dispersion, (2) moderate data dispersion, and (3) narrow data dispersion.

Figure 10-5 shows a normal distribution curve labeled to show the mean (μ), and the 1σ, 2σ, and 3σ zones around the mean. The decimal figures beneath the curve in Figure 10-5 show the fraction of the total area occupied by that region of the curve. These translate to the following percentages:

μ to 1σ: 34.134% of total area

1σ to 2σ: 13.591% of total area

2σ to 3σ: 2.14% of total area

Summing these for both sides of the mean leads us to the conclusion that 68.28% of values lay in the region $\pm 1\sigma$, 95.45% are within $\pm 2\sigma$, and 99.73% are within $\pm 3\sigma$.

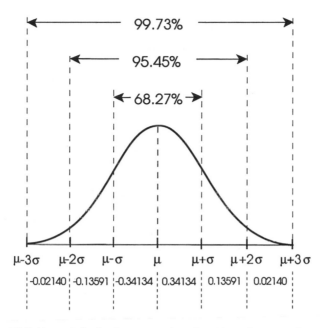

Figure 10-5. Normal distribution curve showing ±1σ, ±2σ, and ±3σ points.

Estimating σ from s

The sample standard deviation can be used to approximate the mean and standard deviation of the population. There will always be some error, but sometimes the estimation is all that we have to work from.

EXAMPLE

The 5-North nursing station in Mytown General Hospital has a "full census"; that is, all 36 beds are occupied. Table 10-1 shows the temperatures of all 36 patients from the afternoon "vital signs" check. Because there are too many data points (36) to reproduce the problem here, I used a computer program (see Appendix B) to calculate the mean temperature, the range end points, the variance (σ^2), and the standard deviation (σ). These values are: mean = 100.4 degrees; range is 96.1 to 106 degrees;

the variance is 7.93 and the standard deviation σ = 2.82. Using a random number generator program I selected five samples from Table 10-1 to make an estimation of the mean, variance, and standard deviation.

TABLE 10-1

Patient Temperatures

103.2	102.5	97.9	100.2	98.7	98.4	104.3	97.2	97.8
105.5	100.9	99.1	96.1	96.8	98.7	98.6	102.1	97.9
105.0	105.0	98.7	103.9	98.6	98.6	100.5	99.4	98.2
101.2	104.2	96.4	106.0	98.6	102.6	103.0	98.7	98.6

1. Calculate the mean of the sample

$$\bar{x} = \frac{98.7 + 102.1 + 101.2 + 98.6 + 104.2}{5}$$

$$\bar{x} = \frac{504.8}{5} = 100.96 \text{ degrees}$$

2. Calculate the variance:

$$s^2 = \left(\frac{1}{5-1}\right) [(98.7 - 100.96)^2 + (102.1 - 100.96)^2 +$$

$$(101.2 - 100.96)^2 + (98.6 - 100.96)^2 + (104.2 - 100.96)^2]$$

$$s^2 = \left(\frac{1}{4}\right)[(-2.26)^2 + (1.14)^2 + (0.24)^2 + (-2.36)^2 + (3.24)^2]$$

$$s^2 = \left(\frac{1}{4}\right) [5.12 + 1.3 + 0.058 + 5.57 + 10.5]$$

$$s^2 = \frac{22.55}{4} = 5.64$$

3. Calculate the standard deviation

$$s = \sqrt{s^2} = \sqrt{5.64} = 2.37$$

The sample average of 100.96 degrees estimated the 100.4 degrees actual average temperature, while the sample standard deviation of 2.37 estimates the actual standard deviation of 2.82. If the sample is larger, the estimates converge on the actual values for the entire population.

The Standardized Normal Distribution Curve

It's convenient to standardize the distribution curve so that it becomes easy to glean information concerning the nature of the data collected. Consider Figure 10-6. This curve is a normal distribution, with $\pm\sigma$, $\pm2\sigma$, and $\pm3\sigma$ points shown. Because this curve is symmetrical, the mean (μ) and median are the same, and the average deviation is zero. The probability of any x being above μ (i.e., $x > \mu$) is 0.50; likewise the probability of $x < \mu$ is also 0.50.

The normal distribution curve of Figure 10-6 is divided into two areas, labeled P and Q. If the total area under the curve is taken to be the unit area (i.e., total area = 1), then

$$P + Q = 1.00$$

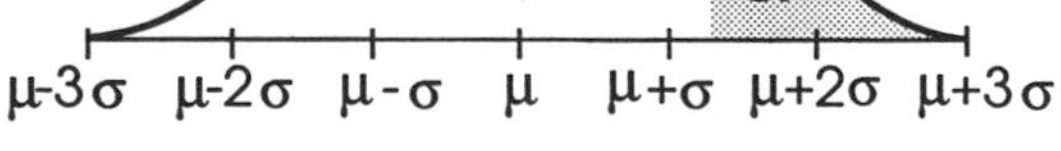

Figure 10-6. Standardized normal distribution curve.

We can designate any value within the range of the curve by using the z-statistic, which is a measure of its distance from the mean, and is computed from:

$$z = \frac{x_i - \mu}{\sigma}$$

or, if the data comes from a small sample of the overall population:

$$z = \frac{x_i - \bar{x}}{\dfrac{\sigma}{\sqrt{n}}}$$

If z is positive, then the value of x is greater than μ, and if z is negative, then x is less than μ. The value of any given x can be denoted by either

$$x = \mu + z\sigma$$

or,

$$x = \mu - z\sigma$$

Because the normal distribution curve plots probabilities of occurrence, knowledge of z is associated with a probability number. If $\mu \pm z\sigma$ is the PQ boundary, then the probabilities can be stated as:

$$P = Pr\{x_i < \mu + z\sigma\}$$

$$Q = Pr\{x_i \geq \mu + z\sigma\}$$

We can look up either $P(z)$ or $Q(z)$ in a statistical table in order to determine the probability of any given x being above or below that value. Consider Table 10-2. Values of z in 0.1 unit increments are arrayed along the left column, while increments of 0.01 units are along the top row. In reading values from this table take your cues from both horizontal (0.01 steps) and vertical (0.1 steps). For example, to find the value $Q(z)$ associated with $z = 1.45$, we look at the number as if it were "1.4 + 0.05," so look down the left column to "1.4" and along the top row to "0.05." The cell where the row and column intersect contains the value $Q(z)$ for z = 1.45, which is 0.0735. We can calculate P from this number because

$$P = 1 - Q$$

Table 10-2

Areas under the Standardized Normal Curve

z	.00	.01	.02	.03	.04	.05	.06	.07	.08	.09
0.0	.5000	.4960	.4920	.4880	.4840	.4801	.4761	.4721	.4681	.4641
0.1	.4602	.4562	.4522	.4483	.4443	.4404	.4364	.4325	.4286	.4247
0.2	.4207	.4168	.4129	.4090	.4052	.4013	.3974	.3936	.3897	.3859
0.3	.3821	.3783	.3745	.3707	.3669	.3632	.3594	.3557	.3520	.3483
0.4	.3446	.3409	.3372	.3336	.3300	.3264	.3228	.3192	.3156	.3121
0.5	.3085	.3050	.3015	.2981	.2946	.2912	.2877	.2843	.2810	.2776
0.6	.2743	.2709	.2676	.2643	.2611	.2578	.2546	.2514	.2483	.2451
0.7	.2420	.2389	.2358	.2327	.2296	.2266	.2236	.2206	.2177	.2148
0.8	.2119	.2090	.2061	.2033	.2005	.1977	.1949	.1922	.1894	.1867
0.9	.1841	.1814	.1788	.1762	.1736	.1711	.1685	.1660	.1635	.1611
1.0	.1587	.1562	.1539	.1515	.1492	.1469	.1446	.1423	.1401	.1379
1.1	.1357	.1335	.1314	.1292	.1271	.1251	.1230	.1210	.1190	.1170
1.2	.1151	.1131	.1112	.1093	.1075	.1056	.1038	.1020	.1003	.0985
1.3	.0968	.0951	.0934	.0918	.0901	.0885	.0869	.0853	.0838	.0823
1.4	.0808	.0793	.0778	.0764	.0749	.0735	.0722	.0708	.0694	.0681
1.5	.0668	.0655	.0643	.0630	.0618	.0606	.0594	.0582	.0571	.0559
1.6	.0548	.0537	.0526	.0516	.0505	.0495	.0485	.0475	.0465	.0455
1.7	.0446	.0436	.0427	.0418	.0409	.0401	.0392	.0384	.0375	.0367
1.8	.0359	.0352	.0344	.0336	.0329	.0322	.0314	.0307	.0301	.0294
1.9	.0287	.0281	.0274	.0268	.0262	.0256	.0250	.0244	.0239	.0233
2.0	.0228	.0222	.0217	.0212	.0207	.0202	.0197	.0192	.0188	.0183
2.1	.0179	.0174	.0170	.0166	.0162	.0158	.0154	.0150	.0146	.0143
2.2	.0139	.0136	.0132	.0129	.0125	.0122	.0119	.0116	.0113	.0110
2.3	.0107	.0104	.0102	.0099	.0096	.0094	.0091	.0089	.0087	.0084
2.4	.0082	.0080	.0078	.0075	.0073	.0071	.0069	.0068	.0066	.0064
2.5	.0062	.0060	.0059	.0057	.0055	.0054	.0052	.0051	.0049	.0048
2.6	.0047	.0045	.0044	.0043	.0041	.0040	.0039	.0038	.0037	.0036
2.7	.0035	.0034	.0033	.0032	.0031	.0030	.0029	.0028	.0027	.0026
2.8	.0026	.0025	.0024	.0023	.0023	.0022	.0021	.0021	.0020	.0019
2.9	.0019	.0018	.0017	.0016	.0016	.0016	.0015	.0015	.0014	.0014
3.0	.0013	.0013	.0013	.0012	.0012	.0011	.0011	.0011	.0010	.0010

For the case of $Q = 0.0735$, therefore, $P = 1 - 0.0735 = 0.9265$. We know, therefore, with $z = 1.45$, for any given value x_i the probability of it being greater than $\mu + 1.45\sigma$ is 0.0735, and the probability of it being less than $\mu + 1.45\sigma$ is 0.9265.

The z-statistic and the standardized normal distribution curve can be used in scientific experiments to verify or reject a hypothesis. Let's look at several examples.

EXAMPLE 1

A college nutritionist was studying male college freshmen. One of his factors was the weight of incoming freshmen in September, which he compared with their weights in May. A total of 986 freshmen were weighed during registration (much to the perplexity and consternation of the Registrar—scientists do that to people sometimes). Their individual weights varied from 98 lbs to 356 lbs, with a mean of $\mu = 156$ lbs and a standard deviation $\sigma = 16$. What is the probability of a randomly selected male freshman being selected who weighs more than 200 lbs?

Solution:

$$z = \frac{x_i - \mu}{\sigma}$$

$$z = \frac{200 - 156}{16} = \frac{44}{16} = 2.75$$

Looking up $Q(z)$ for $z = 2.75$ shows a value of 0.0030, from which we deduce that the probability of randomly choosing a male freshman who weighs over 200 lbs is 0.003, or 0.3%. Conversely, the probability of the random student weighing less than 200 lbs is $P = 1 - Q = (1 - 0.0030) = 0.997$.

EXAMPLE 2

A site in the southeast United States, dating from the colonial period, was excavated by a team of archaeologists. They uncovered a large amount of glassware shards, and wanted to determine whether or not the glassware came from a local maker. A large amount of glassware from that maker was known and had been catalogued. One of the properties of the local glassware was that it contained the heavy metal lead in the proportion 0.79 mg of lead (Pb) per 1000 grams (1 kg) of glass (0.79 mgPb/kg) with a standard deviation (σ) of 0.058.

Fifteen samples were taken to the university's chemistry lab to have the lead content measured. The results were: 0.91, 0.92, 0.94, 0.92, 0.92, 0.93, 0.90, 0.96, 0.92, 0.93, 0.92, 0.91, 0.93, 0.92, 0.94. These data provided a standard deviation (s) of 0.015. The archaeologist wanted to know the likelihood that this sample came from a population with a mean of 0.79 mgPb/kg and a standard deviation of 0.058, so calculated the value of z:

$$z = \frac{0.924 - 0.79}{0.058} = \frac{0.134}{0.058} = 2.31$$

From the table, with $z = 2.31$, $Q(z) = 0.0104$ and $P(z) = 1 - Q = 0.9896$. Thus, only about 1% of the samples from the local glassmaker would fall into this region, so the hypothesis is rejected.

Note that it is not impossible for the glass to have come from the local maker, but it is highly unlikely. Unless there was some special cause for the sample to be very unusual, we have to conclude that the sample came from another source.

EXAMPLE 3

A factory has six ovens used for the curing of adhesives. These ovens are designed to provide a set temperature between 500 to 1000 degrees Fahrenheit for periods of 1 to 10 hours. When the ovens were originally installed last year, a large number of temperature measurements were taken on each oven. In each case, the set temperature was 750°F. Oven number 1 produced an average temperature of 778°F over 240 trials, with a standard deviation of 2.7.

Recently, manufacturing complained that the oven was out of control, and was the cause of a lot of rejected work. A process engineer measured the actual temperature as 785°F. What is the probability that this temperature would occur in an oven that is under control?

$$z = \frac{T - \mu}{\sigma}$$

$$z = \frac{785 - 778}{2.7} = \frac{7}{2.7} = 2.59$$

With $z = 2.59$, the value of $Q(z) = 0.0048$ and $P = 1 - Q = 0.9952$. It is unlikely that the set temperature would've occurred if the oven were in control. But, being a good engineer, she also knew that one data point, even one so far out on the normal distribution curve, does not make a case. Over the next ten days there were fifteen uses of the oven scheduled, so she measured the temperature each time. The mean turned out to be 820°F with a standard deviation of 11.2. The oven was very much out of control.

The lesson to learn from this example is that a single data point can point the way, and may lead to truth, but is insufficient for making a determination. Multiple trials are needed before any statement of significance is possible.

Testing for Normal Distribution

The normal distribution curve is so widely found that one is often tempted to assigned it to almost any set of data. Sometimes we know that's not right because the rules for binomial distribution are seen to be in effect. In other cases, however, before we automatically assign "normalcy" to any given data set it might be prudent to perform a little test called the *quantile graph*. The quantile plots the standardized normal distribution curve along the vertical axis, and the x_i data along the horizontal axis (Figure 10-7). If the data are normally distributed, then the data points will lay along a straight line.

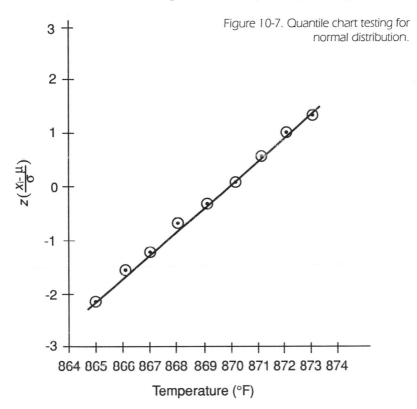

Figure 10-7. Quantile chart testing for normal distribution.

The t-Distribution and Student's t-Test

It is often the case that σ cannot be known for certain, even though s can be calculated. In those cases σ must be estimated. One method for doing that neat trick is the t-distribution. In 1908, W.S. Gosset, who worked for the Guinness brewery, developed the t-distribution, but was unable to publish it under his own name because of restrictions from his employer. They consented to the use of a pseudonym, however, and Gosset chose the name "Student."

The t-distribution is close to the normal distribution (Figure 10-8), the difference between the two being determined by the number of samples available. The t-statistic is an estimate of the standard deviation, and is used similarly to the z-statistic discussed earlier:

$$t = \frac{\overline{x}_i - \mu}{\left(\dfrac{s}{\sqrt{n}}\right)}$$

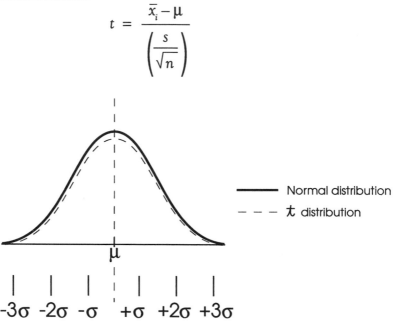

Figure 10-8. The t-distribution curve.

The dispersion of the t-distribution is a function of both the number of samples and the number of parameters (such as σ) that are estimated. When $n \to \infty$, however, $t \to z$ and the two distributions become identical.

The number of parameters that must be estimated sets the "degrees of freedom" of the t-distribution, which is the number of observations less the number of parameters that must be estimated. For example, when only the standard deviation of the population is estimated, then the number of degrees of freedom is one less than the sample number. For example, if 10 samples are taken, then the number of degrees of freedom is $10 - 1$, or 9. Table 10-3 shows a t-distribution that is used in a manner similar to Table 10-2.

Table 10-3

Students *t*-Distribution

Degrees of Freedom	.01	.05	.01
1	3.078	6.314	31.821
2	1.886	2.920	6.965
3	1.638	2.353	4.541
4	1.533	2.132	3.747
5	1.476	2.015	3.365
6	1.440	1.943	3.143
7	1.415	1.895	2.998
8	1.397	1.860	2.896
9	1.383	1.833	2.821
10	1.372	1.812	2.764
11	1.363	1.796	2.718
12	1.356	1.782	2.681
13	1.350	1.771	2.650
14	1.345	1.761	2.624
15	1.341	1.753	2.602
16	1.337	1.746	2.583
17	1.333	1.740	2.567
18	1.330	1.734	2.552
19	1.328	1.729	2.539
20	1.325	1.725	2.528
21	1.323	1.721	2.518
22	1.321	1.717	2.508
23	1.319	1.714	2.500
24	1.318	1.711	2.492
25	1.316	1.708	2.485
26	1.315	1.706	2.479
27	1.314	1.703	2.473
28	1.313	1.701	2.467
29	1.311	1.699	2.462
30	1.310	1.697	2.457
60	1.296	1.671	2.390
120	1.290	1.661	2.358
∞	1.282	1.645	2.326

Other Distributions

The normal distribution curve shows up everywhere in science and technology, so much so in fact that it is a reasonable choice when making *tentative* assumptions of the distribution of an unknown data set. But not all data sets are normally distributed, so you must be cautious when attributing normalness to a data set that may in fact be non-normal.

Figure 10-9 shows several possible alternatives to the normal distribution set that are, in fact, found in real data sets. Figure 10-9a and 10-9b show data sets in which the frequency of occurrence is skewed in one direction or the other. In Figure 10-9a we see negatively skewed data, while in Figure 10-9b we see a positively skewed data set.

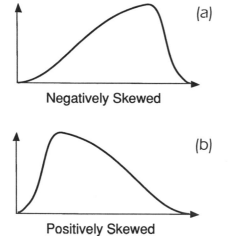

(a)

Negatively Skewed

Figure 10-9. Skewed distributions: (a) negatively skewed; (b) positively skewed; (c) bimodal distribution.

(b)

Positively Skewed

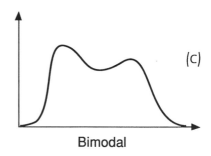

(c)

Bimodal

A measure of the skewedness of the data set can be seen in the skewness coefficient (γ), which is calculated from:

$$\gamma = \frac{\sum\limits_{i=1}^{n} (x_i - \mu)^3}{\sigma^3}$$

This equation works because it evaluates to zero when the data are distributed equally about the mean, but will give a negative result for data sets such as Figure 10-9a, and a positive result for data sets such as Figure 10-9b.

Two alternate estimations of skewness are called Pearson's coefficient of skewness, which are calculated from:

$$\text{Skewness} = \frac{\overline{x}_i - x_{\text{mode}}}{\sigma}$$

or,

$$\text{Skewness} = \frac{3(\overline{x}_i - x_{\text{median}})}{\sigma}$$

The skewness sometimes identifies data that is not normally distributed, so allows you to use alternate techniques. It also helps you spot problems in method or equipment when data that should be normally distributed shows up heavily skewed. An example showed up in a medical laboratory instrument used to measure blood sodium and potassium levels. The machine had become contaminated from previous samples (a frequent happening). It was noticed because quality control records showed a markedly skewed data set over the course of several hours.

The bimodal distribution seen in Figure 10-9c is another non-normal data distribution that sometimes shows up. It is often caused by having mixed populations, with overlapping means.

Binomial Distribution

The binomial distribution is very common. Three requirements are necessary for the binomial distribution:

1. There are only two possible outcomes for each trial. An example is the flip of a coin: Only *heads* and *tails* are possible as outcomes. . . unless you are me (but more of that later).

2. There is a fixed number of trials, each of which are totally independent of the others (again the flip of a coin is the example).

3. The probability for each of the outcomes is the same for each trial. If the probability is 0.50 for both possibilities, then it must be 0.50 at every trial.

The binomial distribution sometimes looks very much like the normal distribution curve, especially if the number of data points (n) is large and the probability for each trial is close to 0.5.

For the binomial distribution, the variance and standard deviation formulas can be made much simpler than the formulas given earlier in this chapter. They are:

$$\sigma^2 = np(1-p)$$

and,

$$\sigma = \sqrt{np(1-p)}$$

where n = the fixed number of trials

p = the probability of success in one of the n trials

$(1-p)$ = the probability of failure in one of the n trials.

I alluded to a coin that, for me, was not binomial. It really happened. At Old Dominion University several of us were sitting lazily in Webb Center (the student union) and debating whether or not to go to math class. One of the guys said: "Let's flip a coin—heads we go to the beach, tails we go to the King's Head Inn [a local student beer joint on the edge of campus]... and if it stands on end we'll go to class." He flipped the coin, it dropped out of his hand when he caught it, rolled around on the carpeted floor, and came to rest against my chair...on end. We went to class.

11

MAKING SENSE OF YOUR DATA—REGRESSION ANALYSIS

DATA FROM measurements are often recorded during experiments, and then plotted on a scatter diagram such as that shown in Figure 11-1. The horizontal, or x, axis is used for the *independent variable*, while the vertical, or y, axis is used for the *dependent* variable. The latter may be labeled y, or f(x), unless some specific quantity is being measured. For example, you might graph a pressure as a function of time. The x-axis would then be labeled in units of time (t), while the y-axis would be labeled P, P(t), or P = f(t).

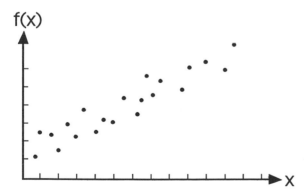

Figure 11-1 Scatter diagram plots data points as f(x) vs. x.

Alternatively, suppose an electrical heating element is immersed in a vessel containing water, and you suspect that the temperature of the water is a function of the electrical current

167

flowing in the heater element. You might plot current (I) along the x-axis, and temperature (T) along the y-axis, and label it either T or $T = f(I)$.

When all of the data points are plotted, a pattern may emerge that suggests some relationship. That is, it appears there is some *correlation* between x and $f(x)$. In Figure 11-1, for example, it appears that the values of $f(x)$ increase with increases in the value of x. This fact implies that there is some correlation between x and $f(x)$. Because $f(x)$ increases as x increases, we say there is a positive correlation between x and $f(x)$.

Figure 11-2 shows several different types of correlation that might be found in data sets. A strong positive correlation is found in the data of Figure 11-2a, while approximately the same degree of negative correlation (i.e., $f(x)$ decreases as x increases) is shown in Figure 11-2b. A weak-to-moderate positive correlation exists in the data of Figure 11-2c, indicating either that the connection between them is not strong, or possibly that some measurement error exists or uncontrolled variable is at work. When there is no correlation (Figure 11-2d), then there is little or no evidence of a pattern in the placement of the data points.

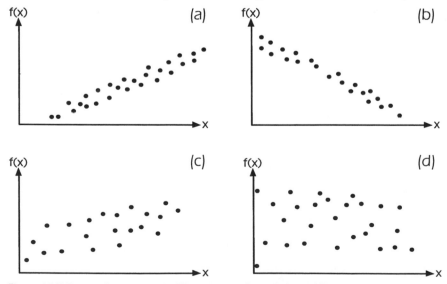

Figure 11-2 Scatter diagrams with different types of correlation: (a) Strong positive correlation; (b) strong negative correlation; (c) weak positive correlation; (d) poor or no correlation.

It is tempting to assume that a causal relationship exists between x and $f(x)$ when we see a strong correlation between the data in a scatter chart. Sometimes such an assumption is valid, and sometimes it is not. For example, it is reasonable to assume that the temperature of the water in which an electrical heating element is immersed is a function of the current flowing in the heater because we know the physics of electrical heaters in water baths. At other times, however, a strong correlation may indicate nothing at all about causality, only that the two factors seem to have a relationship. They may be related only through a third factor. For example, suppose we plot the number of dance studios in our town against the number of new cars sold, and note a strong positive correlation. Is it reasonable to assume that either new car sales cause dance studios, or that dance studios cause new car sales? Not at all, but rather both of these factors may well be related to the fact that the town population doubled over the last ten years, or that a new plant at the edge of town brought a wave of new employment opportunities and prosperity that permits residents to have nice things like late model cars and dance lessons.

For purposes of analysis we can make the assumption that x causes $f(x)$, provided that we understand the limits of this assumption with respect to reality. From this assumption we can construct a *mathematical model* of the relationship between x and $f(x)$ that predicts values of $f(x)$ for values of x other than those that were actually measured. The strength of our model is revealed in the reliability of that prediction for data outside the measured set. It matters little whether or not a causal relationship exists *if knowledge of x predicts $f(x)$.*

What is meant by "mathematical model" in this context is a line or curve that best fits the data. Figure 11-3 shows the naive way: connect the dots. Unfortunately, like many a simple solution this one is easy to do, looks elegant, but is utterly wrong. Just connecting the dots in no way predicts the values of $f(x)$ for values of x that are not in the previously measured data set.

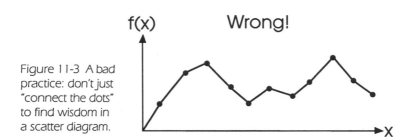

Various models can be pressed into service, and which to use depends on the nature of the phenomenon being investigated. For example, if we know the phenomenon is likely to be a quadratic equation, then we may want to use a model such as Figure 11-4a. In other cases, we might use an exponential decay or rise, both of which are shown in Figure 11-4b. In most cases, when prior knowledge is lacking, we make the assumption that the data is *linear*, that is, a straight line (Figure 11-5a) of the form:

$$y = f(x) = a + bx$$

Where:

    a is the y-axis intercept point
    b is the slope of the line.

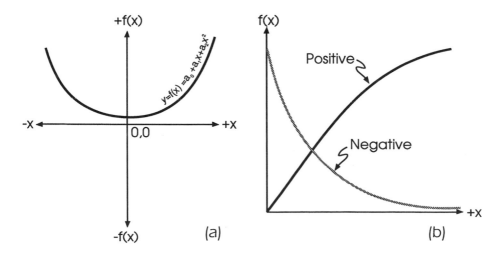

Figure 11-4 (a) The quadratic function; (b) positive and negative exponential functions.

When faced with a collection of data, you could simply take a straightedge and draw a line through the center of the data, attempting to get as many points as possible on the line, or, failing that, as many on one side of the line as on the other. But this solution is also wrong, and does not come as near the mark as is possible with better technique.

A better way is to calculate a straight line (that is, find the slope and y-intercept) that concisely summarizes the relationship between the two variables. This is called the regression line, or the *least-squares* line. The assumption made when constructing such a line is that the relationship is linear, and that the y component of each data point is subject to a certain small error ε, that is:

$$y = \hat{y} = a + bx + \varepsilon$$

[NOTE: $\hat{y}$ *means estimate of* y]

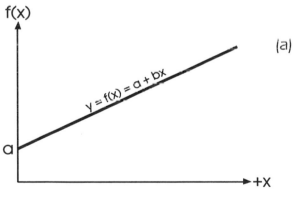

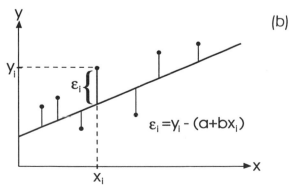

Figure 11-5 (a) Linear function of the form y = f(x) = a + bx; (b) A linear regression line to describe data in a scatter diagram. The error terms ε_i are the difference between the regression line and the actual value y_i. The line is constructed to minimize the distance between itself and the data points.

The goal of the least-squares fit is to minimize the distance between each point and the selected line (Figure 11-5b), making each ε term as small as possible. The equation for finding b is:

$$b = \frac{\sum\limits_{i=1}^{n} x_i\, y_i - \sum\limits_{i=1}^{n} x_i \sum\limits_{i=1}^{n} y_i}{n \sum\limits_{i=1}^{n} x_i - \left(\sum\limits_{i=1}^{n} x_i\right)}$$

Since the regression line always passes through the centroid $(\bar{x}, \bar{y})$, we can find a by:

$$a = \bar{y} - b\bar{x}$$

which is the familiar:

$$a = \frac{\sum\limits_{i=1}^{n} y_i}{n} - b\, \frac{\sum\limits_{i=1}^{n} x_i}{n}$$

where n = the number of pairs of data points.

The *correlation coefficient* r is a number between 0 and 1, where 1 indicates an exact fit between the line and the data, and 0 indicates no fit at all.

$$r = \frac{n \sum\limits_{i=1}^{n} x_i\, y_i - \sum\limits_{i=1}^{n} x_i \sum\limits_{i=1}^{n} y_i}{(n-1)\, s'_x\, s'_x}$$

$$s'_x = \sqrt{\frac{s_x^2}{n-1}}$$

$$s'_y = \sqrt{\frac{s_y^2}{n-1}}$$

EXAMPLE

Consider the function $y = 2 + 4x$. A data set on this equation is created with the aid of a random-number generator on a handheld calculator in order to simulate experimental data. The data are shown in Table 11-1. (I'm indebted to Art Ciarcowski of the Food and Drug Administration for this example and the data table.)

x	y
6.6	20.1
9.1	45.0
17.4	67.4
17.9	73.4
12.9	95.9
13.3	98.9
13.6	103.1
18.1	90.6
29.1	92.7
25.6	114.7
36.5	119.0
35.0	109.8
30.0	121.8
30.7	117.3
39.3	145.8
47.5	171.6
40.9	174.8
48.5	206.5

Table 11-1

$$s_x^2 = 2{,}956.23$$

$$s_y^2 = 35{,}382.28$$

$$s_{xy}^2 = 9{,}128.47$$

$$s_y^2 - s_x^2 = 32{,}426.05$$

$$b = \frac{32{,}426.05 + \sqrt{(32{,}426.05)^2 + 4(9{,}128.47)^2}}{(2)\,(9{,}128.47)}$$

$$b = 3.81$$

$$a = \frac{1{,}968.40}{18} - \frac{(3.81)\,(472)}{18}$$

$$a = 9.33$$

The line is therefore defined by:

$$y = 9.33 + 3.81x$$

and the coefficient of correlation is:

$$r = \frac{9{,}128.47}{\sqrt{(2{,}956.23)\,(35{,}382.28)}} = 0.89$$

Orthogonal Least Squares

An implicit assumption of the least-squares method is that all of the error terms are in one axis. But there are situations where the linear expression is not $f(x) = a + bx + \varepsilon$, but more nearly resembles $y \pm \varepsilon_y = a + bx + \varepsilon_x$. This permits independent errors in both x and y axes. Such a situation can be found in a sensor calibration curve where the sensor outputs a voltage that is proportional to some parameter, such as pressure. To calibrate, we use a manometer to measure "actual" pressure (but with error) and a voltmeter to measure the sensor output value (with its own error). Both errors are independent of sensor error, but the x-axis voltage error can be combined with sensor error. A means for dealing with these is *orthogonal least squares*.

Orthogonal least squares also works well when the data has error in only one axis (which might normally suggest the use of conventional least squares), but in which the available data points are highly clustered in only a small region of the range.

The goal in calculating the slope (b) and y-axis intercept (a) in orthogonal least squares is to minimize the distance between the fit line and the data points, along lines *orthogonal* to the fit line (Figure 11-6). The calculations are as follows:

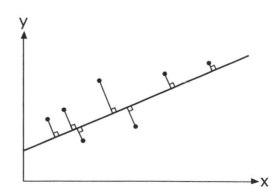

Figure 11-6 Orthogonal linear regression line minimizes the orthogonal distance between the line and each data point.

Slope b:

$$b = \frac{(s_y^2 - s_x^2) + \sqrt{(s_y^2 - s_x^2) + 4(s_{xy})^2}}{2 s_{xy}}$$

y-intercept a:

$$a = \frac{\sum_{i=1}^{n} y_i}{n} - b \frac{\sum_{i=1}^{n} x_i}{n}$$

Correlation coefficient r:

$$r = \frac{s_{xy}}{\sqrt{s_y^2 s_x^2}}$$

Where:

$$s_x^2 = \sum_{i=1}^{n} x_i^2 - \frac{\left(\sum_{i=1}^{n} x_i\right)^2}{n}$$

$$s_y^2 = \sum_{i=1}^{n} y_i^2 - \frac{\left(\sum_{i=1}^{n} y_i\right)^2}{n}$$

$$s_{xy} = \sum_{i=1}^{n} x_i y_i - \frac{\sum_{i=1}^{n} x_i \sum_{i=1}^{n} y_i}{n}$$

When the same data are used in the orthogonal model as were used in the conventional model before, the line is described by $f(x) = 28.38 + 3.09x$, as opposed to $f(x) = 9.33 + 3.81x$, with the same correlation coefficient of 0.89; compare with the original expression of $f(x) = 2 + 4x$.

Using Linear Regression Models

Both of these forms of linear regression are, in essence, mathematical models of the data set. We rely on these models to make sense of the world measured by our instruments. But over-reliance on a single model can lead you to disastrous consequences in evaluating data. Several errors commonly crop up in using regression equations to make predictions:

- Don't use a regression equation to make predictions if no significant linear correlation exists. This one seems obvious, but is often not heeded!

- If the population is different from that from which the sample data were gathered, then no predictions can be made. For example, if you use the average salaries of males to develop a regression equation relating salaries to education level, those results don't necessarily work for females.

- Always stay within the reasonable range of the sample data when making predictions. For instance, it would be ridiculous to use a regression equation relating the birth weight and height of babies to predict the height of a baby weighing 100 lbs!

The assumption underlying linear regression is, not surprisingly, that the relationship between the independent and dependent variable is linear; that is, that it follows the $y = f(x) = a + bx$ relationship. But there are other legitimate relationships as well as linear. For example, the *quadratic model* can be used for data that fits the model:

$$y = a_1 x^2 + a_2 x + a_0$$

Or, the data may be a closer fit to an exponential function, or a polynomial of several orders higher than the quadratic equation. It is quite easy to erroneously select the easy linear model, especially when dealing with phenomena about which you have no prior knowledge. Sometimes, you get "hunches" from a knowledge of the subject matter and the data that the proper curve is other than linear. Other times, the "lay" of the data points on the scatter diagram may make the suggestion for you. Other times there are suspicious biases in the lay of the data points once the linear regression line is calculated. Let's consider an example.

Table 11-2 shows data collected in an experiment, while Figure 11-7a shows the same data

X	Y
0	0.5
1	0.7
2	1.1
3	1.7
4	2.5
5	3.5
6	4.7
7	6.1
8	7.7
9	9.5
10	11.5

Table 11-2

plotted on a scatter diagram. Although a practiced eye will see a familiar general shape to the line connecting these points, it is all too easy to "see" a straight-line relationship with some data points above the line and others below it—exactly what linear regression is designed to produce. Figure 11-7b shows the

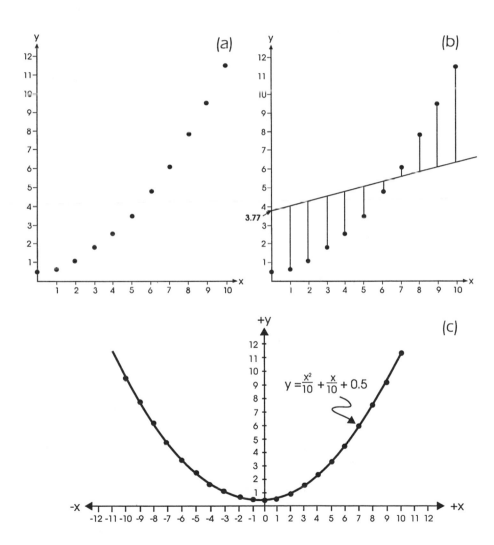

Figure 11-7 (a) Plot of some data...is it a candidate for linear regression? (b) Linear regression line-fit for the data of (a). Note the symmetry of the data points (suspicious!); (a) A quadratic regression would have been a better fit to the data!

same data set with the linear regression line plotted through the data. A suspicious bias has appeared. All the points $x_i < 7$ are below the line, while points $x_i > 7$ are above the line. Further, the length of the residuals lines between the point and the least squares lines uniformly decreases as $x \to 7$, and then just as uniformly increases as x increases above 7. For a process that is supposedly due to random variation in either the measurement process or the phenomenon being studied, these two facts are strikingly regular—in fact, suspiciously so. The truth is revealed when more data points from the same set are plotted (Figure 11-7c), and we discover that they fit a quadratic shape. The parabola is evident in this graph, but was less so in Figure 11-7a.

The lesson? You must be very careful when doing data fitting. You must not become too enamored of the model, remembering all along that the goal is to find a truth about the data.

Calibrating

Linear regression methods are sometimes used to create calibration curves for scientific instruments, to find out how much the instrument deviates from the ideal and to generate correction factors. Suppose an electronic pressure sensor is used to measure pressures in the 0 to 120 Torr range. Also suppose that the pressure has an output function that relates the output voltage (V_o) of 10 mV/Torr. If a pressure manometer is used to measure the same standard pressure as the sensor (Figure 11-8a), and both pressure and output voltage data points taken, we can do a linear regression to determine the actual calibration curve (Figure 11-8b).

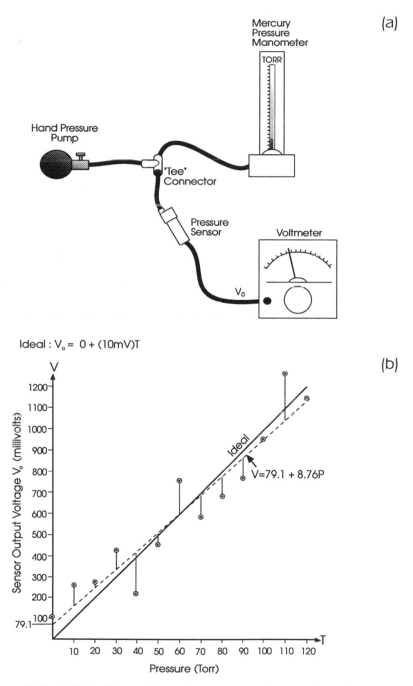

Figure 11-8 (a) Calibrating an electronic pressure transducer against a mercury manometer; (b) Plot of data and its linear regression.

Reference

Mario F. Triola, *Elementary Statistics, Fourth Edition*, The Benjamin/Cummings Publishing Co., Inc., Redwood City, California, 1989.

You Can't Test Them All– Sampling Data

T HE PURPOSE of collecting scientific data is to gain insight on some class of events or things in nature. These events (or things) could be the occurrence of a certain characteristic among white mice, a defect in the final product of a production process, a positive reaction to a drug, or any of millions of different things that can occur. The entire class of all possible observations, including all possible members, is called the *population* (or in some contexts, the *universe*). When all members of a population are examined, the data collection is called a *census*. When only a few randomly selected representative members are examined, the data are said to be *sampled*.

While it is easy to argue that the best data collection is that which is painstakingly taken on the entire population (census), that assertion is not necessarily correct, especially when the population is a very large number. In many cases, it is simply not possible to examine each member of the population in order to determine the incidence of the characteristic under study. Do you examine all 62 milliard (one milliard equals 10^9) blue crabs born in the Chesapeake Bay last year to see how many of them have some trait associated with pesticide run-off from agriculture? For such a study, a *sample* of the total crab population is used. Sampling is a statistical means for inferring truth about a large population by examining only a small portion of that population.

Or how about people like political pollsters, research psychologists, or sociologists? They may wish to study the attributes of a large group of people, and it would be impossible to examine them all. For these researchers, sampling the population is the only practical way to proceed.

In other cases, it is wasteful and unproductive to examine an entire population. Suppose we are producing 17,000 nylon automobile door lock handles every day, and need to keep track of our injection molding process. If the key parameter in this quality test is the weight of the handle, then we could weigh all 17,000 to make sure that each is within tolerance of, say, ±4.5 grams. Not very smart, especially when Megapollutor Motors wants those door knob handles today! Instead of measuring the weight of every handle, we would instead take a small sample (five to ten) of the knobs every hour, and weigh them. This procedure will show any trend that develops, which might indicate a problem with the process, procedure, or materials used. In some factories, entire lots of product may be accepted or rejected on the basis of a small sample taken at random during the process of manufacture.

In still other cases, testing each and every individual within the population is just plain stupid. Suppose you are making 5.56 mm ammunition for the Army's M-16A2 main battle rifle. You could, I suppose, test fire every single bullet made, but that defeats the purpose of making bullets: What would you issue to the troops? (Of course, if you are an antiwar activist, then this type of testing protocol might have a certain amusing logic of its own.)

Besides the inconvenience of testing all members of a large population, there is a secondary effect that is sometimes a problem. If too many individuals must be tested, especially over a long period of time, then several types of error can crop up. These errors are based on the assumption that testing the entire population takes a long time.

Fatigue error comes about when the individuals doing the testing and measuring become tired or bored with the task. They may become sloppy in performing their tasks, and consequently mistake rates and magnitudes increase.

A case in point is a study done on clinical medical laboratories. In a test procedure done by manual (as opposed to automatic) means, the error rate was found to be about 5% (clinically acceptable in that context). But when the results were refined enough to see the hourly error rate, it was found that the error varied significantly both by shift and by the hour within the shift. It was found that day and evening shifts were extremely busy, so errors tended to creep up as technicians hurried through the task. The 11pm-to-7am "graveyard shift" (bad name in a hospital!) tended to produce the best results because they had the lightest load. On the hourly rates within a shift, it was found that the daytime (7am-to-3:30pm) errors were nearly zero in the first hour (when technicians were fresh) and nearly 25% in the last hour (when the techs were fatigued). A 25% error is considered clinically absurd by medical people.

Drift error occurs when an instrument or procedure is used over a long period of time to make a measurement. Instruments drift (or "age" as they say), and require periodic recalibration. The recalibration procedure, in itself, however, sometimes changes the error enough for a nicely trending plot of data to become useless.

Another form of long-term drift occurs in some manual test procedures when there is a nearly imperceptible change in the way the experimenter carries out the measurement over time. Such a change might be difficult to spot, especially if the procedure seemed to be carried out correctly. In one medical study, physicians were tasked to inject a bolus of a chilled intravenous solution (D_5W for those who *must* know) into the patients being studied. The test was a measure of cardiac output (liters of blood per minute) using the *thermodilution method*. It was found that minute, seemingly inconsequential, differences in the length of time the physician held the syringe changed the temperature of the solution enough to raise the error in the measurement beyond acceptable limits. Initially, the injection time was much longer than later on (when the physicians were more comfortable with the procedure), so all the early data was useless.

Between-tester variation. Like it or not, different people tend to get different results when performing the same supposedly objective test. This error is not too terrible if trending is done, or if there are a number of experimenters whose results are averaged together in some reasonable manner. But when a lone tester is used for a long period of time, and is then replaced, the new person's error may be different enough to corrupt the data in the long run.

Accumulation-of-error error. Errors can be propagated in some types of test, especially if there is a drift error in the process. But even without a drift error, there is a possibility of accumulation of error when the results of today's tests are compared with earlier tests. "Running averages" or "sliding window" data sometimes exhibits this problem. In these cases, you must discard the earliest data (which may contain higher or lower error terms), and use only the most recent data. For example, one neurology researcher in a study of evoked potentials EEG brain waves always used the last 100 subjects to keep track of his process. That is, after 250 patients, his data would be on the 150th through 250th. But error terms in the earlier samples were used to calculate values for the 150th, which also contained its own error, so the overall error was greater than normal.

Certain measurements made in science are often treated as if they were sample data, even though the "population" might be an obscure concept in that particular case. For example, when the blood pressure of a test subject is measured several times in a row, such as might be done when calibrating instruments or in an experimental process, then the resulting data are treated as if the measurement was a sample of a large population. See Chapter 8 on measurements for more details.

A *sample* is a small portion taken from the entire population. If the sample items are randomly selected, and if every member of the population stands an equal chance of being selected (both rules are important), then it can be reasonably inferred that the attributes of the sample apply to the population as a whole. A statistical analysis is used to make assertions about the whole population from sampled data. Conversely, you can do a probability analysis to home in on the likely nature of the samples before they are taken.

Randomness

Sampling only works properly when the sample is taken randomly. A sample that is truly random is called a *fair sample*, just as in card games a fair deck is one that is complete and well-shuffled. Two of the rules for randomization were mentioned above, but bear repeating (they are *very* important).

1. The samples were selected according to a random process.

2. All members of the population had an equal chance of being selected.

To these, we can add a third rule:

3. The chances of any one individual being selected does not depend on prior selections.

The first rule is obeyed by inventing a procedure whereby the selection is made randomly. For example, a well-mixed pot of bingo chips could be drawn from the mixing bowl by a blindfolded person. Unfortunately, this procedure doesn't always work. In the first draft lottery of the Vietnam era, critics claimed a certain bias to the selections because men born in December had a disproportionate chance of being drafted. The December date cards had been dropped into the hopper last, so were too close to the surface in the insufficiently mixed barrel. And it was very important which dates were drawn early, for the low numbers were drafted first and had a high probability of going into mortal combat. (I still recall that first draft lottery: A young man fainted in the student center of Old Dominion University, while watching the lottery on TV, because his lottery number was 3.) It seems that the 365 numbered tags were not well enough mixed to ensure that true randomness occurred. This type of bias is called the *data collection bias*. The following year, the draft lottery was conducted in a more random manner—the Selective Service Office asked the National Bureau of Standards to help design a random-selection process.

Another randomization method might be to select numbered members of the population using a table of random numbers, or the random number generator on a computer or handheld calculator. This method works very well when done correctly, but there are some potential problems. These problems are based on the fact that there is no such thing as a truly random random number generator. The random number generator on my MS-DOS IBM-clone computer is an example. If I execute the RND command in BASIC, then it will return a pseudorandom number between 0 and 1. By proper arithmetic, this random number can be translated to any desired range.

But there is a fly in its electronic ointment. The random number generator has to be given a seed number to do its thing (which is true of most pseudorandom sequence generators). If you know the pattern of seeds, then you can predict the pattern of random numbers. In fact, some low-tech encryption methods dating back to the 1930s depend on this fact. To the outside observer who lacks knowledge of the seed number, the pattern appears random. But to those in the know....

Many computer programs used for science experimentation use a random number method that automatically reseeds the random number generator every time it's called by using the computer's built-in timer/clock function (see sidebar). If too little time passes between successive random numbers (it only takes a few tenths of a second), then the randomness is destroyed...and a dishearteningly unrandom pattern can appear. In a BASIC program that I wrote to teach experimentation, called *Ovengame*, the goal is to find the right oven temperature to cure adhesives (Appendix C). The player sets the temperature, but the program adds or subtracts a random number from the set temperature to yield the actual oven temperature. In order to ensure randomness, I placed a logo on the screen "ADHESIVE COOKING Please be patient" for a number of seconds that is proportional to the set curing time. This feature is touted to "add realism" to the simulated experiment, but it was actually done to reseed the computer's random number generator properly. On one oven, without the delay, the standard deviation on a set temperature of 750 degrees averaged 4, but with the delay the average standard deviation went to 28.

Even if the random number is correctly taken, there can be problems. One of my college professors, in a psych course I believe, told us of a researcher who was trying to discern some truth about soldiers. The local army post commander was cooperative, and allowed her to select test "victims" from

---

° Generating Random Numbers in BASIC

The BASIC program shown below will generate a random number from 0 to a maximum value (N) that you specify. It uses the computer's internal clock to reseed the random number generator. If you hit "RUN" (i.e., F2 on MS-DOS machines) twice in quick succession, the same or adjacent numbers (such as 82 82 or 56 57) will consistently appear, yet waiting a couple of seconds produces widely differing random numbers most of the time. This illustrates why a time delay is needed in Ovenmaster.

```
100 REM Random Number
    Generator
110 N = <fill in highest number
    desired in range>
120 A%=VAL (MID$ (TIME$, 7, 2))
130 RANDOMIZE A%
140 A = INT (RND*(N+1))
150 PRINT A
```

among his men. Four companies of about 200 men each were arrayed in parade formation so that she could select her subjects. She punched the random number key on her handheld calculator, and ".37" popped up on the display. So, she deftly counted off every 37th man... and wound up with a sample of all senior sergeants! Did the random number generator fail? No it didn't, but there was a bias in the population. *Not every member of the population had an equal chance of being selected.* Remember Rule 2 above? Why did this happen? Her population was *arrayed in military formation*, such that squads, platoons, and companies were standing in the same repetitive rank order. That means that the noncommissioned officer ranks repeated across the formation... every 37th soldier was the sergeant in charge of a 36-man platoon!

The type of sampling done by the researcher above is called *systematic sampling*. It works well in many situations, and is a reasonable tool to use. It was not capriciousness that caused her to use a random number to make the selection of participants. The problem came because of an orderliness in the population—i.e., the military formation—that made a random number a source of error rather than strength. She would have obtained a lot better random sample if she had adopted a different method, perhaps calling up a different random number between 1 and 37 for each platoon. With a sufficiently large sample she would've more than likely wound up with some of each rank... which is what she wanted in the first place.

Actually, the researcher made a second mistake as well. If she wanted to study "typical" soldiers, then she should have used more than one Army post for her population. Army posts tend to be specialized: Infantry at Fort Gordon, armored units at Fort Hood, and airborne troops at Fort Benning. There might be some special attribute (height? weight? cussedness?) among the types of men selected for each branch of the Army, and these factors could possibly bias her results.

Another example of bias is seen in a case that has become a staple for teachers of statistics and/or research methods. It is called "The 1936 *Literary Digest* Presidential Poll" problem. In 1936, President Franklin Delano Roosevelt was running for reelection against lackluster Republican Alf Landon. Outside Kansas, where he'd been governor, not many people even knew his name. *Literary Digest* magazine mailed out 10 million questionnaires, of which about 2.4 million were returned. From these, a sample was taken and the magazine analyzed the results. They predicted 57% of the vote for Landon, and 43% for FDR. It even seemed to make sense, for the Great Depression was in full force, and any politico could be expected to feel a little heat from the electorate. When the election was held, however, the actual results were FDR 62% and Landon 38%. Fatally embarrassing for the *Literary Digest* pollsters!

What happened? Why didn't sampling work this time? It seems that the population of possible voters was taken from nearly every *telephone directory* in the country. But there was the fly in the ointment! It was the Great Depression, and an awful lot of people didn't have telephones, even when they had them prior to the catastrophic economic downturn. In addition, telephones were not nearly as widespread then as now; many rural dwellers didn't even have phone lines to their homes by 1936, never mind phone service. As a result, it was the economically well-off who were polled, and this group would include an unusually high number of Republicans. In addition, this group probably contained a higher proportion of well-educated, white, city-dwelling, white-collar workers, who were politically active. . . or at least politically interested.

Another mistake made by the *Literary Digest* pollsters is that they ignored the nonresponders. If 2.4 million people responded out of 10 million polled, then 7.6 million people (76%) failed to return the questionnaire, so didn't even have a chance of being selected in the sample. In this group there was probably a larger proportion of the lesser educated, the non-white, the blue-collar workers, the rural dwellers, and the politically inert.

If you had polled the college where I spent my freshman year about the Vietnam war, the results would've been difficult to extend to the population of college kids in general. That school was filled to the brim with recently discharged military (especially Navy), and other older students whose attitude was: "I did my duty, so why shouldn't you?" When a Students for a Democratic Society (the infamous SDS) speaker (some say "Rabble Rouser") left town following the May Day "demonstration" over the Cambodia invasion, he was quoted on local TV calling the school a "...*hotbed of political rest.*" Compare with other campuses that May of 1968!

Whenever a situation exists where such ill-fitting data is found, there is said to be a bias in the sample. One such bias is seen in the examples above: systematic exclusion of some portions of the population from the possibility of selection. That is the so-called *Systematic Bias*—for example, excluding those who couldn't afford, didn't want, or couldn't get, telephone service. Another type of bias is also seen above: The *Nonresponse Bias*. There may be good reasons why the nonresponders didn't respond, and it can seriously affect the outcome of the study.

Another source of bias is the *Shenanigans Bias*, also called the *Monkey Business Bias* or *Dirty Tricks Bias*. This type of bias comes from something like "stuffing the ballot box." In political polls, for example, some wild fanatics for one cause or another will "drive all night through a snowstorm" (as one radio talk show guest put it) in order to be polled. There was once a poll to see if the citizens of a Canadian city wanted to go to the expense of creating a new sports stadium. The poll showed a strong sentiment for the project. Later it was discovered that members of a local athletic team, their families, friends and supporters had...shall we say..."stuffed the ballot box." They mailed in 286 of the 510 responses received.

Congressional aides are amused when the same letter arrives from a large number of constituents. Letters are their "sampled data." And the perpetrators of mail-in campaigns no longer need to distribute printed copies for people to sign and send. They simply load the letter onto a sympathetic computer bulletin board, and dozens of people can make their own "original" copy. Of course, no one is fooled by this ploy. When four dozen pro or con letters arrive, all have exactly the same wording and format, despite appearing to each be originals. Or *are* they fooled!

Still another type of bias is the *Wrong Venue Bias*, in which the population from which samples are drawn is not truly the real population. For example, suppose you wanted to measure the physical stamina of typical college students, and you selected your sample from the cross country team or the Bike Hiking Club. Would it be a fair sample of the universe of college students? Probably not.

Sampling techniques are often used in acceptance testing of materials and products that are bought in large numbers. For example, electronic integrated circuits ("chips") and machine screws tend to be purchased in large quantities by manufacturing organizations. Part of the specification in the purchase order will be an *acceptance number* (AN) or *acceptable quality level* (AQL) that states the number of defectives permitted in each batch. Sometimes, the AN/AQL will be specified in percentages (for example, 2% defective), while at other times it is specified in terms of numbers defective per hundred or per thousand items (e.g., 2/1000). Each batch of material is sampled by the incoming quality inspectors. If the sample shows more defectives than the permitted AN/AQL, then the entire lot is rejected.

Modern management theory, put forth by Deming and others, rejects the notion of AN or AQL. The Japanese typically don't use this concept; to them AQL = 0 is the acceptable norm.

The difference between the American and Japanese approach is seen in an order for dynamic random access memory chips (DRAM), the heart and soul of any competent computer, ordered by an American electronics manufacturer from a Japanese semiconductor supplier. They had an AQL of 10 bad chips per pallet of 10,000 chips (the standard order size for them). The Japanese packed the defective chips in a small bag and sent it along with the order and a note: "We don't understand why your business needs defective chips, but here they are. In order to prevent confusion with the good chips we've packed the defective chips separately in this bag."

Or consider the case of a government contractor manufacturing mercury-wetted electrical relays for the Department of Defense. The in-plant inspector caught them replacing two good relays in the packed shipment of 100 with two that had failed tests. "Whadya doing?" asked the Fed. "Giving you what you ordered—2% AQL . . . these are your two bad relays."

These stories may be apochryphal, as "industry-wide" stories tend to be, but they serve to illustrate a principal fallacy in using statistical technique to bad end. The original purpose of the AN/AQL idea was to use statistics to ensure minimum defectives, even though recognizing (incorrectly, it turns out) that perfection is not possible. Modern managers ask: "Why not zero percent AN?"

Sampling Arithmetic

It is common practice when doing sampling to take the mean average of the sample, and then use it as a representation of the entire population. The usual convention for sampled data (x_i) is to use the English letter x with a bar over it for mean, and s for standard deviation. The equivalent for the mean of the entire population is designated by the Greek letter μ, and the standard deviation by the Greek letter σ. The mean of n observations of x is the arithmetic average:

$$\overline{x}_s = \frac{x_1 + x_2 + x_3 + \ldots + x_n}{n}$$

If the sample is properly taken, with due regard for random-ness, then the sample mean is a reasonable estimate of the population mean.

The standard deviation of the sample is similar to the standard deviation for the population, but the denominator is $n - 1$ rather than n:

$$s = \sqrt{\frac{\Sigma (x_i - \overline{x})^2}{n - 1}}$$

The sample means form a normal distribution curve, even when the population is not normally distributed. The standard deviation of the sample mean can be calculated, and it forms a standard error term:

$$S_{sm} = \frac{S_x}{\sqrt{n}}$$

which is a good estimate of the population equivalent:

$$\sigma_{sm} = \frac{\sigma_x}{\sqrt{n}}$$

The standard deviation of the sample mean is used as an indicator of the error surrounding a measure of x:

$$x \approx \overline{x} = \frac{\sigma_{sm}}{\sqrt{n}}$$

EXAMPLE

A certain pressure sensor is designed to have an output factor of ten millivolts per Torr (10 mV/Torr) over a range of −100 to +500 Torr. A standard test condition places a 100-Torr pressure, as measured by a laboratory grade manometer, on the sensor and measures the output voltage. Long experience, with many measurements recorded and the statistics calculated, indicates that this sensor has a mean of 1072.7 mV at 100 Torr with a standard deviation of 13.4 Torr on a range of 1051 mV/ 100-Torr to 1100 mV/100-Torr.

A sample of seven measurements ($n = 7$) is taken with the following values resulting: 1055 mV, 1059 mV, 1051 mV, 1099 mV, 1089 mV, 1059 mV, and 1053 mV. Calculating the mean and standard deviation shows that $x = 1066.4$ mV, $s = 19.3$. Calculate the standard deviation of sample σ_{sm} and express the error terms for the mean.

$$\sigma_{sm} = \frac{\sigma_x}{\sqrt{n}}$$

$$\sigma_{sm} = \frac{13.4 \text{ Torr}}{\sqrt{7}}$$

$$\sigma_{sm} = \frac{13.4 \text{ Torr}}{2.65} = 5.06$$

The measurement error is ±5.06 Torr, so the expression of the pressure measurement is:

$$x = 1066.4 \pm 5.06 \text{ Torr}$$

We can also compute an estimate of the standard error knowing only the range and the sample size ($n \leq 15$).

$$S.E._{(est.)} = \frac{x_{max} - x_{min}}{n}$$

For example, in the example given above the standard error is the value of σ_{sm}, or ±5.06 Torr. The estimate given by the equation above is:

$$S.E._{(est.)} = \frac{1099 - 1051}{2.65} = \frac{48}{7} = 6.86$$

This figure is often within ±20% of the correct value, and is sometimes used as an "on-the-fly" estimate in the lab. The error of the example above is around 35%, but it would have dropped if the sample size were larger. The estimate is a good "in-process" indicator, but should never be used to replace the correct form in the final product.

For small populations, where the sample size is a substantial proportion of the total population, we sometimes make a correction of the standard error using the form:

$$\sigma_{sm} = \frac{\sigma}{\sqrt{N}} \times \sqrt{\frac{N-n}{N-1}}$$

Where:

N is the population size
n is the sample size
$n > 0.1N$

Confidence Intervals

When we make a measurement using sampling techniques we are estimating the expected value μ using the mean of the sample. We can calculate the probability of μ being close to the mean using

$$Probability = \left(\frac{x - k\sigma}{\sqrt{N}} < \mu < \frac{x + k\sigma}{\sqrt{N}} \right)$$

The value of the parameter k is:

P	k
0.90	1.64
0.95	1.96
0.98	2.33
0.99	2.58

We can use these probabilities to make statements about the true value of the population mean from the sample mean, or to predict that the probability of a sample about to be extracted will have a k percent probability of being no further than $k\sigma/\sqrt{N}$. For the various confidence levels:

90% confidence:

$$\sigma_{sm} = \frac{1.64\,\sigma_x}{\sqrt{N}}$$

95% confidence:

$$\sigma_{sm} = \frac{1.96\,\sigma_x}{\sqrt{N}}$$

98% confidence:

$$\sigma_{sm} = \frac{2.33\,\sigma_x}{\sqrt{N}}$$

99% confidence:

$$\sigma_{sm} = \frac{2.58\,\sigma_x}{\sqrt{N}}$$

Figure 12-1a shows the situation for fifteen trials when the standard deviation of the population is known (although its mean, μ, may not be known). All intervals are equal length because the standard deviation of the population does not vary. If the population standard deviation (σ) is not known, however, the estimated value (s) will vary from one trial to another, so the interval also varies (Figure 12-1b).

The meaning of the vertical range arrows at each trial is shown in more detail in Figure 12-1c. Because each trial consists of a series of sample measurements, it will have a range (minimum to maximum), and its own mean (x) that may or may not be identical to the population mean (μ). Given a sufficiently large, randomly selected sample, the values of the individual measurements within the sample will be approximately normally distributed, as shown in the curve in the inset to Figure 12-1c.

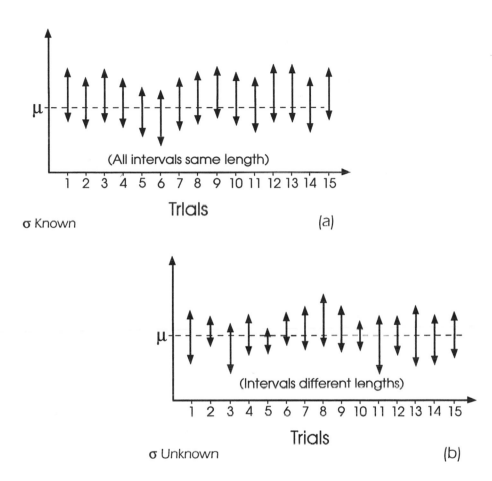

(All intervals same length)

1 2 3 4 5 6 7 8 9 10 11 12 13 14 15

Trials

σ Known

(a)

μ

(Intervals different lengths)

1 2 3 4 5 6 7 8 9 10 11 12 13 14 15

Trials

σ Unknown

(b)

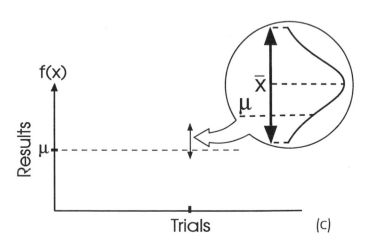

f(x)

Results

μ

Trials

(c)

Figure 12-1
(a) Measurements
made in 15 different
trials, showing the
limits of confidence
for each when all
intervals are of
the same length;
(b) Same data for
intervals of differing
length; (c) statistical
meaning of the con-
fidence interval...it's
a normal distribution
around the mea-
surement value.

Figure 12-2 shows the meaning of the confidence intervals in graphical form. For a 90% confidence level (±.64σ), the values have a high probability of falling into the unshaded area of Figure 12-2a. Note well that it is not impossible for a value to fall within the two 5% regions at the tail ends of the normal distribution, it is merely less likely. Similar curves are shown in Figures 12-2b and 12-2c for the 95% confidence level (±1.96σ) and 99% confidence level (±2.58σ).

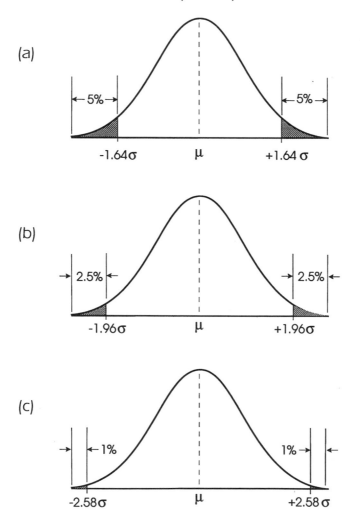

Figure 12-2 The light area indicates the region in which the actual data point exists with (a) 90 percent confidence; (b) 95 percent confidence; (c) 99 percent confidence.

The critical factor in the success of any sampling operation is the size and quality of the sample. It is generally believed that the larger the sample, the better the estimation of the characteristics of the population. But this rule of thumb is not always true, partially for reasons given earlier. For example, sampling 50% of a population using rules that introduce biases is a lot less viable than using a truly random method to select a smaller fraction of the population. Larger sample sizes are valid only if randomly selected according to the rules given earlier, and measured in such a manner as to not introduce any of the errors that sometimes come with very large samples.

In general, actual sample size is somewhat more important than the proportion of the population being sampled. Thus, sampling 1000 out of 1,000,000 is only marginally less accurate than sampling 1000 out of 10,000.

The minimum size of a sample can be determined from the level of confidence desired for the operation. You can solve equations such as the previous group of four for N in order to find a minimum sample size.

If the standard deviation of the population is known with good accuracy, we can calculate the sample size if we can set a desired upper limit for the sample standard deviation, σ_{sm}. If this number is M, then the sample size is found from:

$$N > \frac{\sigma^2}{M^2}$$

In other cases, the standard deviation of the population is not known, but a reasonable estimate is available. Such an estimate might be deduced from the design of the experiment, past history (for example, previous repetitions of the experiment), or other factors. We know that the approximate length of a confidence interval is:

$$L = \frac{2t\hat{\sigma}}{\sqrt{N}}$$

from which we can solve for the minimum value of N:

$$N > \left(\frac{2t\sigma}{L}\right)^2$$

where: L is the confidence interval length, t is the value from a student's t-test table for the number of degrees of freedom, and σ is the estimated standard deviation.

One thing that you can see from these equations is that the improvement in confidence level is proportional to the *square root of the sample size*. As a result, gaining an order of magnitude (10x) improvement in confidence requires a sample that is 100 times larger than the old sample size.

Using And Abusing Graphs

GRAPHS ARE ONE of the most effective ways of presenting scientific and engineering information. The human eye perceives things on graphs that are not obvious in equations or tabulated data. Graphs can quickly show the behavior of a data set, mathematical equation, or other forms of numerical data better than either prose or the mathematical statement of the equation. Graphs can be really beautiful.

Although truth and beauty emerge from graphs, so can deception and lies. While no one would seriously suggest that all erroneous graphs are made with evil intent in mind, there is a sufficiently large amount of such fraudulent activity around that a prudent person is cautious. For this reason, we will also examine some common mistakes—or intentional frauds— associated with graphs.

Many of the problems with graphs arise from the very same factors that are their inherent strengths: The human eye catches facts very rapidly in graphical form. Trends and patterns that are not easily discerned in either tabulated data or equations become immediately visible in many forms of graph. On the other hand, by incorrect use—or fraud—patterns can seem to emerge when none actually exist. In this chapter we will examine some of the basics of graphing data, including both methods of graphing, and the different types of graphs.

Cartesian and Polar Coordinates

Perhaps the most familiar graph is the x-y or *Cartesian coordinate* form (named after Rene Descartes, 1596–1650), shown in Figure 13-1a. This coordinate system is also called *rectangular coordinates*. Two axes are constructed; the vertical axis is labeled "y" and the horizontal axis is labeled "x." When the full range of possible values is used, there will be −x, +x, −y and +y segments of the two axes, with the *origin* (the point of intersection) being given the values of x=0 and y=0.

Any point on the Cartesian coordinate is defined by two points, one each from x and y axes. By convention, these points are listed as an *ordered pair* of the form (x,y). The point P shown in Figure 13-1a is properly specified as (4,3), an ordered pair in which x=4 and y=3.

The four quadrants of the Cartesian system are sometimes labeled *I, II, III* and *IV*, after the fashion shown in Figure 13-1a. The values of x and y in these coordinates follow the pattern in Table 13-1. In many cases, only one quadrant is shown because of the nature of the data (such as Figure 13-1b), but the graph is nonetheless a Cartesian coordinate system even though limited to one quadrant.

Table 13-1

Quadrant	Sign of X,Y Data
I	+X, +Y
II	−X, +Y
III	−X, −Y
IV	+X, −Y

It is common practice to graph independent variables along the x-axis (the "domain"), and dependent variables along the y-axis (the "range"). This idea is seen in Figure 13-1b where the curve is labeled "y = f(x)." Dependent variable y is a function of independent variable x. In other words, knowledge of x predicts the value of y. We are not making a statement of causality here (x does not "cause" y), but rather indicate a sufficiently strong correlation to be able to predict the value of one factor by knowing the value of the other.

Point P in Figure 13-1a can also be defined by a length (R) from the origin (0,0), and an angle (α) from the reference +x axis. In this system, the reference +x axis is used to denote 0 and 360 degree points, with angles measured in the counterclockwise direction. Thus, the +y axis is at 90 degrees, −x axis is at 180 degrees, and the −y axis is at 270 degrees. When angles are given in negative degrees, the rotational direction is clockwise; i.e., −90 degrees and +270 degrees both represent the −y axis.

When the latter system is used, point P can be located by knowing only the length of R and the angle with respect to the reference axis; such coordinates are called *polar coordinates*. For example, the length R is 5, and angle α is 36.9 degrees, so we can describe point P as 5∠36.9. Thus, both (3,4) and 5∠36.9 describe the same point P, but using different notation. Figure 13-1c shows polar coordinates without the cartesian overlay. Converting from one coordinate system to the other is a matter of making a couple of calculations:

Rectangular to Polar Conversion:

$$R = \sqrt{x^2 + y^2}$$

$$a = \tan^{-1}\left(\frac{y}{x}\right) \quad (-180° < a \leq +180°)$$

Polar to Rectangular Conversion:

$$x = R \cos \alpha$$

$$y = R \sin \alpha$$

Often the data shown on the rectangular coordinate system is not discrete data points (like P in Figure 13-1a) but rather a curve that establishes a relationship between x and y (like y = f(x) in Figure 13-1b).

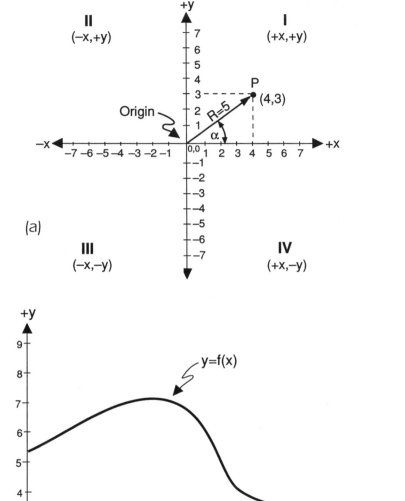

Figure 13-1 (a) Cartesian coordinate system with four quadrants; (b) a single quadrant version often used in practical data recording; (c) polar coordinates; (d) Human blood pressure waveform graphed as a function of time.

In scientific data presentation it is common to find the *x* and *y* labels replaced with other parameters. For example, when measuring the blood pressure of a test subject you might present a Quadrant-*I* graph in which the vertical axis represents pressure, while the horizontal axis represents time.

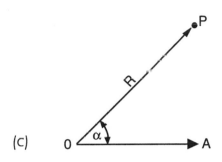

(C)

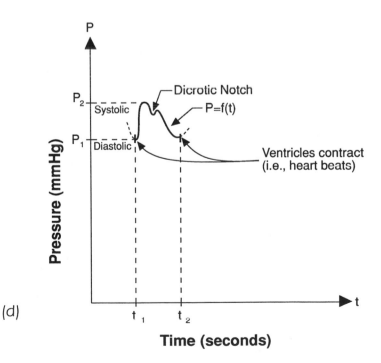

(d)

Time (seconds)

Figure 13-1d shows a representation of the human arterial blood pressure over a single cardiac cycle (only one cycle is shown for sake of simplicity, but the actual waveform is continuous—at least until death). The ventricles of the heart contract, forcing blood into the artery under high pressure. This action causes the pressure to rise from the minimum *diastolic* value (P_1) to a peak *systolic* value (P_2), and then falls back, with a little "glitch" called the *dicrotic notch* that is attributed by some authorities to closing of the aortic valve. The actual pressure curve is a function of time, $P = f(t)$.

A graph such as shown in Figure 13-1d shows more than just the diastolic and systolic values, although those data are also available (indeed, physicians and nurses use such readouts routinely in intensive care medicine). The *shape* of the waveform is sometimes significant, and the physicians can diagnose problems from shape changes. Also, we can find the *mean arterial pressure* (MAP) by integrating, that is, finding the area under $f(t)$ between t_1 and t_2. Counting the number of pulses per minute, or the averaged time between successive pulses, will give the heart rate in beats per minute. A lot of data can be discerned from even a simple graph when it is presented correctly.

Scatter Diagrams

A scatter diagram is used when you suspect that either a "cause and effect" or correlation relationship exists in an experiment. Figure 13-2 shows a typical scatter diagram. The independent variable is mapped along the x-axis, while the dependent variables are mapped along the y-axis. In this case, there appears to be a positive correlation between the x and y values, although this in itself does not prove causality.

One thing to remember when recording data on a scatter diagram: *Do not connect the dots.* A more analytical method is needed to see relationships. See Chapter 11 on linear regres-

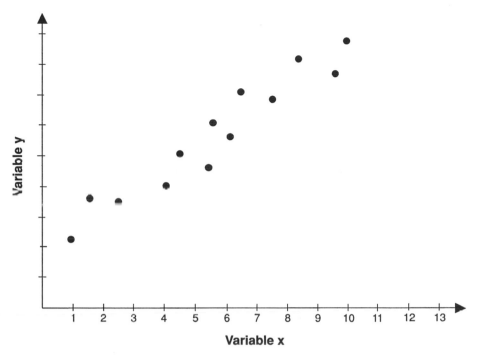

Figure13-2   Scatter diaqram.

sion analysis of data for more information on how to handle these data. Dot-connecting obscures the information content of the scatter diagram.

Run and Control Charts

A *run chart* is used to record sequential data, and is often used in both scientific experiments and industrial process control. Figure 13-3 shows a run chart that records the number of hospital admissions through the emergency room over a two-week period; the dotted line represents the mean average of the data points. Other run charts are used to chart the number of defects in industrial processes, or other factors of interest.

The work of W. Edwards Deming, and Walter A. Shewhart of AT&T, demonstrated the use of such charts to monitor

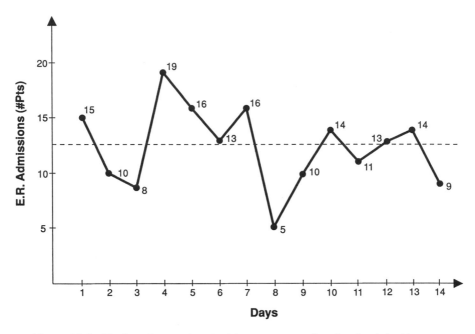

Figure 13-3 Plotting discrete data. In this case, connecting the dots is legitimate because this is a "run chart" and there is no implied relationship between Y and X axes.

industrial processes. In their system, upper and lower control limits (for example, the ±3σ points) are placed on the run chart. When upper and lower control limits are imposed on the run chart, it becomes a *control chart*, and with it you can tell whether or not a process is in control.

Note in Figure 13-3 that the dots between data points are connected. How is this different from a raw scatter diagram? In the raw scatter diagram there is a possibility of there being a relationship between the data points such that $y = f(x)$. That relationship is obscured by connecting the dots, and is far more likely to be modeled by a straight line, a quadratic curve, an exponential curve, or some other curve. In the case where we do not want to connect the dots, the value of y is measured at different values of x through an experiment or observation. The idea is to see how changes in x affect y. On a run chart, which can start out as a scatter diagram, the data points represent discrete samples of the same event over and over. The emergency room admissions above are an example. No one seriously suggests that Friday's E.R. admissions are a function of Thursday's admissions, or that there is some factor that causes the number of admissions to fluctuate day by day.

Another example of a run chart is found in industrial quality control. Suppose a factory is making widgets (all economists use widget factories for examples because it's the only thing they know how to make). A critical parameter, say the diameter of a critical hole, is known to indicate the quality of the product. Because of normal variation, even with the same drill press and drill bit, the diameter of each widget hole varies somewhat from the diameters of other widget holes. In order to measure the process, a fixed sample of widgets is taken from the finished bin at the drill press every hour (e.g., seven widgets are examined hourly); in most cases, the sampled widgets are selected at a random point, but are taken in sequential order (for example, the 57th, 58th, 59th, 60th, 61st, 62nd, and 63rd are selected from the 123 that are produced that hour). The diameters (D) of the critical hole are measured, and the mean average diameter ($\bar{D}$) for that sample, is plotted as a data point on the control chart. At the end of the eight-hour factory shift, eight $\bar{D}$ points will be plotted. These points are often joined by lines in order to spot trends that might indicate machine wear, tool problems, or some other defect in the process that needs to be corrected.

Tally Sheets, Bar Graphs, and Histograms

Tally sheets, bar graphs, and histograms are used to record the relative frequency of occurrence of the specified events or results. There are a number of different ways to record relative frequency data. Suppose we have an event that can have any of eight different results: 0, 1, 2, 3, 4, 5, 6, and 7. We can record them on a check sheet or tally sheet such as Figure 13-4a. Every time the event occurs, a check is placed in the right row, with the fifth in each being a slash bar for ease of counting.

An alternative check sheet method is shown in Figure 13-4b. In this case, an "x" or other symbol is entered every time the event occurs. Notice that the x's pile up in a column form, which suggests a different form of presentation: The *bar graph*. A sample bar graph, for this data, is shown in Figure 13-4c. The use of right-angle axes does not imply that the vertical is a function of the horizontal, but rather indicates only the frequency of the various events.

If we want to normalize the data for relative frequency of occurrence, then we can calculate the proportion of each class as a decimal fraction of the whole. For the data in Figure 13-4c,

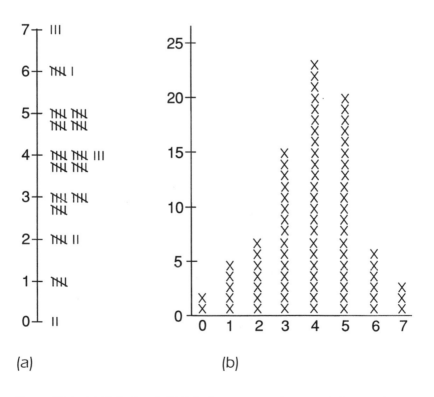

(a)

(b)

Figure 13-4 (a) "Tally sheet"; (b) Tally sheet arranged as a crude bar graph; (c) bar chart; (d) histogram.

the total number of points is 81, so we can calculate the proportion of each event. For example, for event No. 3, the occurrence is 15 out of 81, so the proportion is 15/81, or 0.185. We can display these proportions in the form shown in Figure 13-4d.

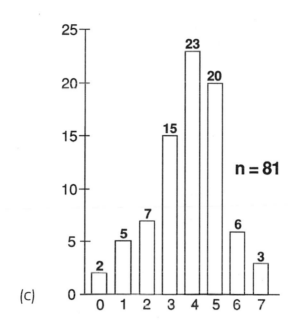

(c)

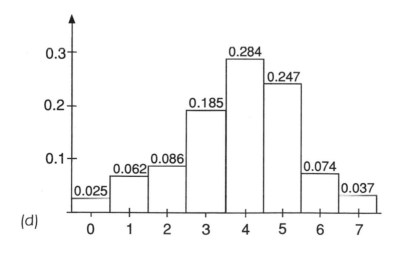

(d)

EXAMPLE 13-1

An adhesives-curing oven is suspected of being out of control because of the irregularity of results being obtained. A very accurate thermometer is mounted in the oven so that the actual temperature can be compared with the set temperature of 750°F. The sorted data obtained over fifteen trials are shown in Table 13-2. It is found that the actual temperatures ranged from 506 to 989°F, with a mean of 784°F and a standard deviation of 157.

Table 13-2

506	517	522	538	541	550	551	583	586	599
626	635	640	664	673	680	697	699	706	728
749	782	785	791	805	835	838	851	851	858
863	875	891	895	899	906	907	938	954	955
956	963	964	965	971	972	973	981	987	989

To construct our chart, we need to find the range ΔR:

$$\Delta R = T_{max} - T_{min} = 989 - 506 = 483$$

We next need to select the number of intervals (K) of temperatures for our chart. It is rarely useful to select as few as three intervals, or more than about twenty (which becomes too busy). For our case, let's use $K = 7$. The selection is sometimes almost arbitrary, but in this contrived case 483 is divisible by 7. The width of each interval is:

$$H = \frac{\Delta R}{K} = \frac{483}{7} = 69$$

A good general rule is to use one more decimal place for setting the intervals than is present in the data. In this case, our thermostat displays temperatures to the nearest degree, so we should use a thermometer that is accurate to tenths of a degree to set the end points. This procedure keeps us from seeing data points on the boundary between two intervals, which would make a counting error.

The lower bound of the graph is either the lowest value in the data set, or the next lowest rounded value below it. In our case, we can use 506°F as the lower bound. We then add the interval width to find the next boundary: 506 + 69 = 575. The next interval starts at 575.1, and so on. These values are shown in Table 13-3:

Table 13-3

INTERVAL No.	RANGE
1	506–575
2	575.1–644.1
3	644.2–713.2
4	713.3–782.3
5	782.4–851.4
6	851.5–920.5
7	920.6–989.6

We can then go to Table 13-2 to count the number of data points in each interval, which are shown in Table 13-4:

Table 13-4

INTERVAL No.	NUMBER IN INTERVAL	PROPORTION
1	7	0.14
2	6	0.12
3	6	0.12
4	3	0.06
5	7	0.14
6	8	0.16
7	13	0.26
	50	1.00

The resultant bar chart is shown in Figure 13-5.

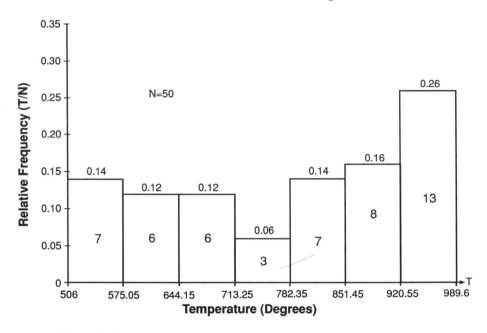

Figure 13-5   Bar chart of oven temperatures.

A histogram is very similar to the bar chart, and indeed in some cases they are the same. But histograms are very different from bar charts in many, perhaps most, cases. In the case of the bar chart, the height of each bar indicates the relative frequency of occurrence of events within the stipulated interval. In a histogram, the area of each bar represents the relative frequency, i.e., the quotient of the frequency and the width of the interval. If it is reasonable to use equal intervals in charting the data set, then it is reasonable to use the bar chart. Indeed, here the bar chart and the histogram are the same. But when it does not make sense to make equal intervals, or where (for whatever reason) it is not either possible or practical, then a histogram makes a better, more accurate, presentation.

The value of the histogram is due to the fact that the human eye sees relative sizes in terms of area, rather than length. Examine the two squares in Figure 13-6a. Square I is exactly twice the size of square II. When originally drawn, square I was 2 inches on each side, while square II was 1 inch on each side. But the eye sees square I as considerably larger than square II even though there is only a 2:1 ratio between the lengths of their sides.

Figure 13-6b shows the same data presented previously in the bar graph of Figure 13-5, but in histogram form. Notice that the geometry is a bit different, and that the effect on the eye is considerably different. In many cases, the histogram confers little benefit, but in other cases the information imparted by the histogram is considerably greater than that of a bar graph of the same data. The differences between histograms and bar graphs will be discussed in a later section when we deal with some common abuses of graphs. (It seems that graphs don't lie, but liars know how to distort graphs.)

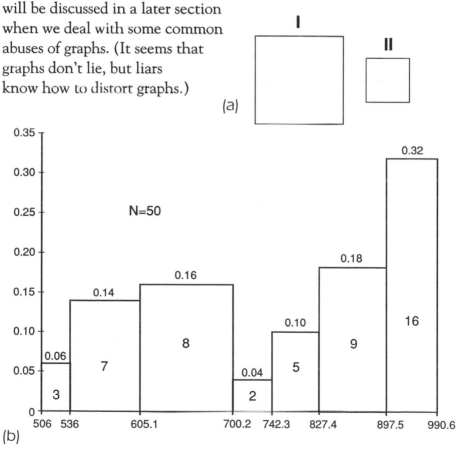

Figure l3-6 (a) The eye sees size in the form of area. Square I is only two times larger than Square II, but looks a lot larger; (b) histogram uses area to show relative frequency.

The range of temperatures along the horizontal axis is very nearly the same as before because the range of the data did not change. A convenient length of graph paper was laid out (in this case, it was chosen to be approximately equal to the horizontal axis of the previous graph). On the graph paper that I used, the length turned out to be 115 small lines, which was then divided by the 484-degree range of data, yielding a step of about 0.24 lines per degree. This factor was then multiplied by the width of each interval to find the number of lines required to represent the interval on the graph paper. When that was done, it became possible to lay out the scale on the horizontal axis.

The vertical axis height (H) is a measure of the relative frequency (i.e., P is the number of measurements in the interval divided by total number of measurements) divided by the width (W) of the interval:

$$H = \frac{P}{W}$$

In the histogram of Figure 13-6b the value of H is placed over the bar for that interval, while the number of measurements that are within the interval are shown inside the bar.

Nonlinear Graphs

So far, all of the graphs that we have seen are linear graphs. But not every process in the universe is amenable to such graphs for one reason or another. A process may be inherently nonlinear, so graphing it on linear charts could result in distortion of relationships. In other cases, the range or domain of possible values is too great for practical graphing on paper smaller than a classroom wall. For example, a certain electronic sensor that detects light produces an electrical current proportional to the light intensity. The normal range of current is a few *picoamperes* (10^{-12} amperes) to about ten *milliamperes* (10^{-2}), or a range of ten orders of magnitude. Showing that current on a linear graph of ordinary size would cause loss of all resolution at the low end, which might be where we are working in low–light-level studies. We could compress the scale into logarithmic form, however, and fit the curve on an ordinary sheet of paper with little effort.

There are two basic forms of graph paper that have logarithmic coordinates: *full logarithmic* and *semilogarithmic*. On full log paper (usually called just "log paper"—is it time to state that most paper is made from logs, even graph paper?) both vertical and horizontal axis are logarithmic (Figure 13-7a), while on semilog paper one axis is logarithmic and the other is linear (Figure 13-7b). In general, semilog paper is used for graphing phenomena that are expressed in the form:

$$f(x) = ae^{bx}$$

while log-log paper is used when the equation is expressed in the form:

$$f(x) = ax^n$$

A logarithmic axis is laid out so that the unit length (L) of the graph is divided into ten divisions that are each a base-10 logarithmic fraction of the total length. The origin of the axis is 1 (not 0), and the highest value is 10. For the numbers 1 to 10 (rounded to the second decimal place):

Number (N)	Log_{10}	%L
1	0	0
2	0.30	30
3	0.48	48
4	0.60	60
5	0.70	70
6	0.78	78
7	0.85	85
8	0.90	90
9	0.95	95
10	1.00	100

Each repetition of the 1 to 10 logarithmic division is called a *cycle*. Figure 13-7c shows a semilog graph in which the vertical axis is divided into two cycles. Here they are labeled 1 to

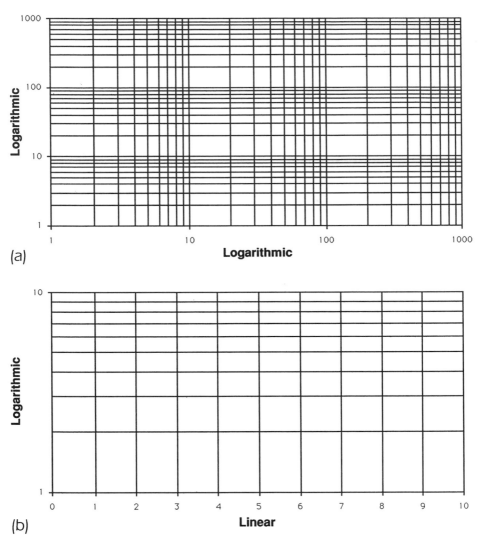

(a)

(b)

10 and 10 to 100, but on some commercial graph paper each cycle will be labeled 1 to 10. Just as likely are fractional labels. I've seen commercial log paper with ten cycles (it was 11x17 paper, not 8.5x11) labelled from 0.001 to 1,000,000. The number of cycles needed depends on the number of decades ("orders of magnitude") of the data that must be plotted.

Let's consider a practical example. Biomedical scientists use special electrodes for acquiring biopotentials from living organisms. Examples include the galvanic (electrical) resistance

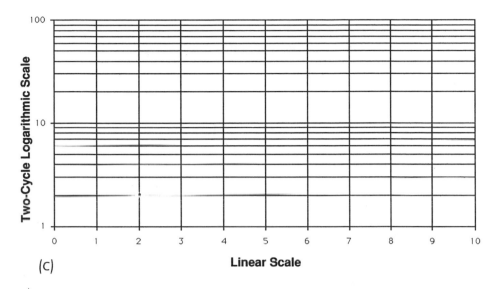

Figure 13-7 (a) Log-log graph scale; (b) semilog graph scale; (c) two-cycle semilog scale.

of plant tissue, the electrocardiogram (ECG) or heart signal, and the electroencephalogram (EEG) or brain wave signal.

Not just any metal will work as a sensor for reasons explained by the chemistry of metals in contact with electrolytes. A key parameter in the design of electrodes for biopotential sensing is the *impedance*, which is similar to electrical resistance to the flow of current (both are measured in ohms). Impedance takes into consideration the resistance, R, which is measured with a direct current (DC), and also certain other opposing factors called *reactances*, symbolized by x. The reactances vary with the frequency of an applied alternating current (AC) signal, so an impedance containing a reactance will also vary with frequency.

Figure 13-8 shows the plot of impedance vs. frequency for a typical bioelectrode over the range of frequencies from 0.1 Hz to 10 kiloHertz (KHz), or five orders of magnitude (10^{-1} to 10^4 Hz); it is plotted on a five-cycle logarithmic axis. Because impedance is within little more than a single order of magnitude, it is plotted on a linear scale from 0 to 14,000 ohms (14 kohms).

The use of the logarithmic scale gives us the freedom to plot five orders of magnitude of data in a small space.

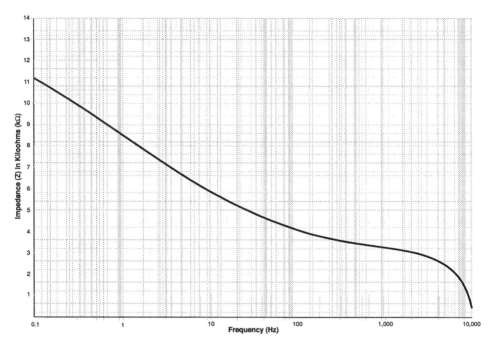

Figure 13-8 Human skin-electrode impedance as a function of frequency shows one use of a semilog graph.

Let's look at log-log and semi-log graphs of the same number in order to see how they are affected. Consider the numbers of the fictional *Zlibarge* scale:

Number Z	LOG Z
10	1
20	1.3
30	1.48
40	1.6
50	1.7
60	1.78
70	1.85
80	1.9
90	1.95
100	2.00

When graphed on linear paper $f(Z) = \log Z$, as in Figure 13-9a, the line is distinctly curved because of the logarithmic nature of the function. When the same is graphed on semilog paper, the curve is very similar (Figure 13-9b). But notice that it straightens out considerably when graphed on log-log paper (Figure 13-9c). These graphs are not terribly practical because they are just meant to show the action of the scale on the shape of the curve.

Figure 13-9
(a) Logarithmic data plotted on linear paper shows a distinct curve; (b) same data on semilog paper; (c) same data on log-log paper (note the straightening).

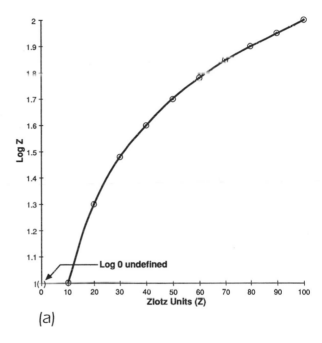

(a)

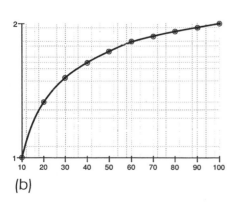

(b)

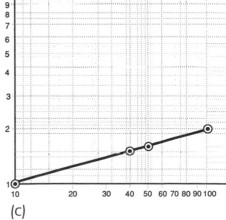

(c)

But the exercise comes in real handy when examining a data set from an experiment. If you don't know the nature of the curve, then you can determine its properties by seeing on which type of paper it graphs to the straightest line. Figure 13-10a

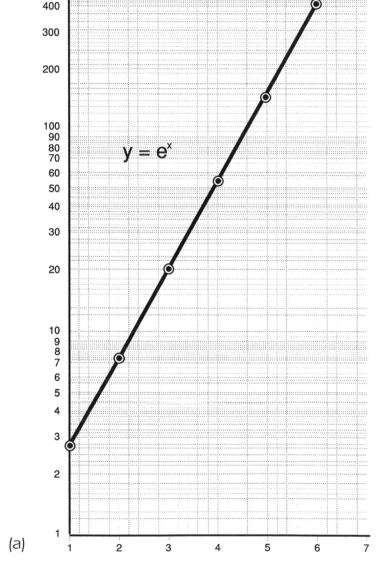

(a)

Figure 13-10 (a) $y = e^x$ data plotted on semilog paper (note straight line); (b) $y = x^2$ data plotted on log-log paper (note straight line).

shows the expression $y = e^x$ plotted on semilog paper. Note that it is a straight line. Similarly, Figure 13-10b shows the expression $y = x^2$ plotted on log-log paper. Again, a straight line is evident.

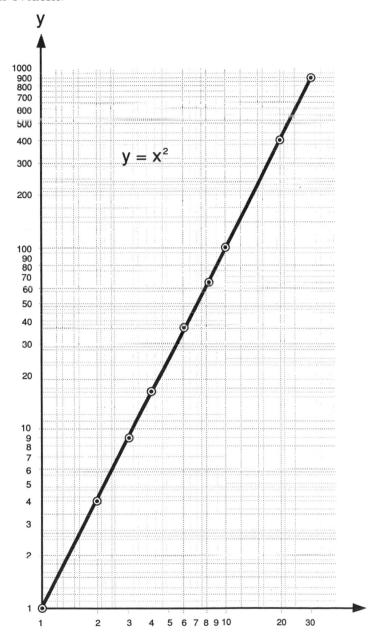

(b)

Integration on a Graph

There are some practical methods for integrating a function that are more or less independent of mathematics training... and those methods work well with experimental data that are not easily fit to some nice, simple equation. Consider a simple function such as Figure 13-11a. The mathematical process of *integration* is the art of finding the *area* under the curve. When the function is simple, like this straight line, then finding the area is easy... just multiply the height and width together, or in the case of this square, it is:

$$Area = (y_1 - 0) \times (T_2 - T_1)$$

or,

$$Area = (15\ units) \times (12\ sec) = 180\ unit\text{-}seconds$$

If the square doesn't start conveniently at zero, like this next one (Figure 13-11b), then we have a different problem, but it's still extremely simple, and the same method still applies:

$$Area = (y_1 - 0\ units) \times (T_3 - T_2\ seconds)$$

$$Area = (15\ units) \times (17 - 5\ sec) = 180\ unit\text{-}sec$$

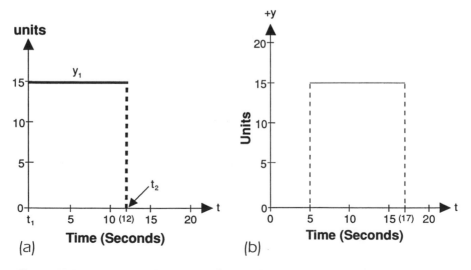

(a)

(b)

Figure 13-11 Integration of a constant function is simple: it is the product of duration and the value: (a) case for time T1 = 0; (b) case for T1 ≠ 0.

But what do you do about more complex shapes? Consider a function such as Figure 13-12a. Suppose you want to know the area under the curve between points *a* and *b*. You can divide this area up into a number of rectangles (Figure 13-12b), and then just add up their respective areas.

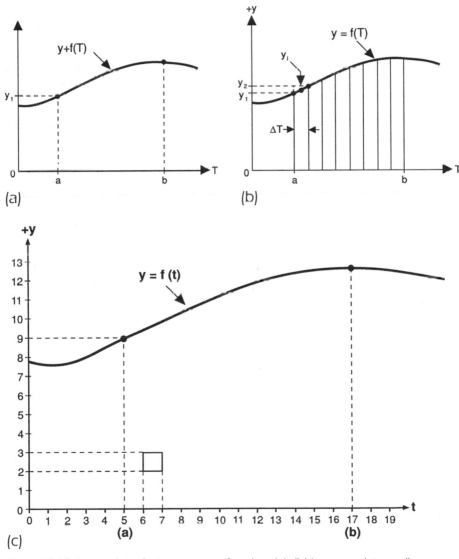

(a)

(b)

(c)

Figure 13-12 Integration of a non-constant function: (a) divide area up into smaller regions, each of which approximates a rectangle; (b) curve for y = f(t) over the time interval a to b; (c) how Fig. 13-12b function can be graphically integrated.

There's an error, because this method is only an approximation. If you make the rectangles small enough, then the error term is minimized. The area of any one rectangle is approximately $y_1(\Delta T)$, although sometimes it is wiser to take the interpolated midpoint between y_1 and y_2, or:

$$y_i = y_1 + \frac{y_2 - y_1}{2}$$

Another neat trick is to use very fine mesh graph paper such as 20x20 divisions per inch. The paper I used was set up with faint lines at 20 per inch, with every fifth line heavier. Draw a one-quadrant graph (Figure 13-12c) with time in seconds along the horizontal axis, and arbitrary units along the y, or vertical axis. By using such graph paper, even a junior high school student can integrate the function by counting squares.

We know the area of the little square because its sides are from 6 to 7 seconds on the horizontal, and from 3 to 4 on the y-axis. Inside this square are 25 little squares. To integrate this function we need only count up the squares. Along the border with the actual curve, many of the large squares are cut into a fraction, and in each case, the fraction can be estimated by counting the little squares. It's a bit tedious, but it gets the job done.

Some Common Abuses

Graphs are a loaded pistol in the hands of the unscrupulous who want to lie without seeming to fib—i.e., they want to convey a false impression even though the data presented is actually accurate. *Horrors!* In other cases, falsehood is presented as fact through inappropriate selection of graphical presentation. You will find these errors in many venues, but when it is dishonest it is probably most frequently seen in sales pitches or viewgraph presentations where the time for proper analysis of the data is not available. Let's take a look at a couple of examples of fibbing with graphs.

I once attended a writer's conference where one of the speakers in the "Economic and Business End of Writing Work-shops was scaring the (mostly self-employed) people in the audience with horror tales of the coming economic bust. The Hunt brothers had been cornering the silver market, and the price of gold was up tremendously. The speaker billed himself as an "international economist" in the conference literature, but he must have flunked "Stats 101."

The speaker had one of those viewgraph machines that allowed him to write on the surface and project his picture or words onto a large screen. He drew the graph in Figure 13-13a, and told the enthralled audience: "I'm no genius, but I know from high school arithmetic that when two points on a graph (using a pointer to indicate "A" and "B" on Figure 13-13a) are such that the latter is higher than the former, the general trend is up—I don't know where it is heading, but the trend is up, up, up." He then launched into a pitch to buy *Krugerands* and other gold bullion investments. Afterwards, in talking to the conference leader, I pointed out an inherent fallacy in his statement. I drew the curve in Figure 13-13b on the speaker's original (still on the viewgraph machine), demonstrating that there may be more than one explanation for "B" being higher than "A."

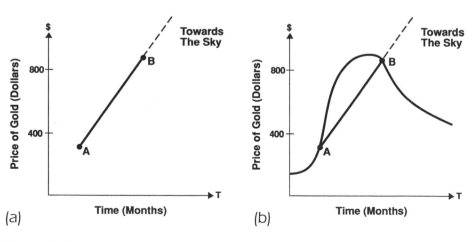

Figure 13-13 Will the price of gold continue upwards? (a) Salesman's prediction; (b) my estimate was nearer the actual price history over the next few months.

The actual price of gold over the next 24 months followed much closer to my curve than the speaker's: It rose rapidly to $800+ per troy ounce, and then dropped back over a few months to around $400 per troy ounce (the price on the day this paragraph was written was $362). *The gold salesman was either ignorant or a liar.*

Let's consider another situation. A production manager is assigned to a factory production line, and desperately needs to prove his worth—especially since he had installed a completely new system of production at this plant. At the end of five quarters he found that production was up from 25,000 units to 30,000 units (an increase of 20%). The graph of production figures is shown in Figure 13-14a, and it reveals a modest success. But the company had invested a large chunk of capital on the manager's word, so "modest" success would be unwelcomed. Thinking deviously, the production manager committed the *suppressed zero fallacy*, and generated the graph of Figure 13-14b.

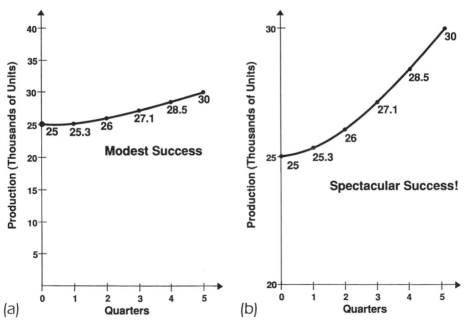

Figure 13-14   (a) Scenario for getting fired — only modest success shows up in the graph; (b) But with the suppressed zero scam the data looks a lot better, <u>huh</u>?

Exactly the same data are shown, but the curve is drastically upwards and tends to give the impression of spectacular succcss! But it's false. Be wary of any graph that doesn't start at zero, or some other natural starting point, unless there is a good reason for the selected bottom reference point.

Another situation is found in the sales department of the same company. A sales program was in effect to move a new line of widgets into a leading position in the marketplace. Management alloted 24 months to the project, and set a goal that sales would be over $50 million per month by the end of the period. During the first few months a very skilled sales manager increased the sales according to the graph in Figure 13-15a. In the tenth month, however, the old sales manager died, went to the Moon, got a new job, or left for some reason, and a new sales manager—a real dud—took over. Sales momentum from the old manager's efforts continued the success for another month or so, and then the results of the new manager's efforts began to filter in. By the end of the program, the sales had fallen off drastically and were well below the projected numbers that management was counting on. The data are tabulated in Table 13-5.

Month	Monthly Sales	Cumulative Sales
1	42	42
2	43	84
3	41	125
4	39	164
5	38	202
6	47	249
7	50	299
8	62	361
9	66	427
10	71	498
11	74	572
12	74	646
13	71	717
14	63	780
15	60	840
16	60	900
17	56	956
18	54	1010
19	49	1059
20	44	1103
21	38	1141
22	36	1177
23	32	1209
24	29	1238

Table 13-5

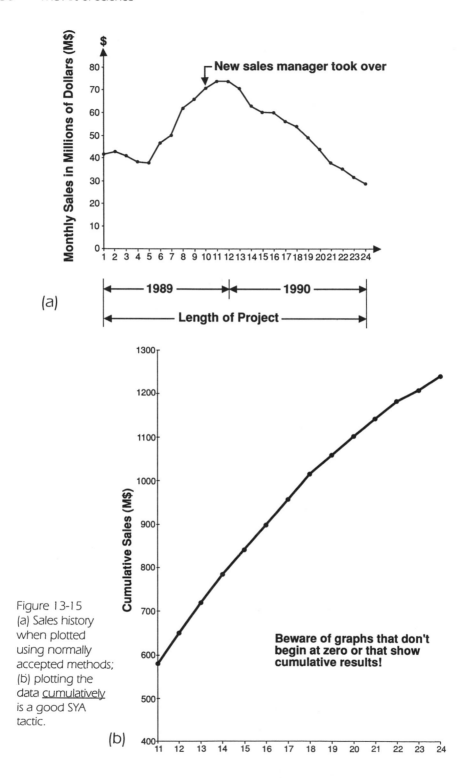

Figure 13-15
(a) Sales history
when plotted
using normally
accepted methods;
(b) plotting the
data <u>cumulatively</u>
is a good SYA
tactic.

How did the sales manager save his job when giving the final report to the company CEO? It's simple: He lied without overtly telling any lies. See Figure 13-15b. This graph shows the cumulative sales, and is thus not terribly sensitive to the reduction of sales per month. Of course, the sharp-eyed accountant in the audience might notice that the slope is tapering off, but without data earlier than month 11 he can't be sure that the change of slope means anything significant.

News organizations sometimes act as if they have their own hidden agenda other than factually reporting the news. (No kidding.) One reason for suspecting such an agenda is that the graphics used present accurate, true data, but in a way that falsifies the impression left on the reader. Consider Figure 13-16a. This graph is based on one actually published in a major weekly news magazine. It presents in graphical form the 18-month U.S. unemployment data shown tabulated in Table 13-6. The impression the reader gets is that unemployment is skyrocketing out of control. Indeed, it may be, but the available data does not support the conclusion. There are at least three graphical sins committed in this presentation.

First, they committed the suppressed zero scam. This graph shows percentage data, so should start at zero percent, rather than five percent.

Second, the vertical divisions are much larger than the horizontal divisions. This type of layout increases the eye's perception of the vertical extent, so accelerates an upward trend more than is justified by the data. In order to stretch the scale, they had to place major divisions every 0.5%, rather than every 1%, as the data implies is correct.

% Unemployed	
January	5.25
February	5.25
March	5.25
April	5.4
May	5.25
June	5.25
July	5.25
August	5.65
September	5.7
October	5.7
November	5.8
December	6.1
January	6.1
February	6.5
March	6.75
April	6.6
May	6.8
June	7

Table 13-6

Third, they carried the trend line off the graph in a rising manner, despite having no data to support the contention that the following month's data would increase. Indeed, the actual data show that there are both abrupt changes of direction and no-change flat regions: There are two abrupt downturns in the data, and six flat zones. The graphically correct way to terminate the graph is to show no trend line following the last month for which data is available. If the point of the graph is to make a prediction, then the known data should be plotted in a solid line, and the prediction in a dotted line (or other format). Changing the type of line lets the reader know that a prediction is being made as to what the trend will probably look like. It is intellectually dishonest to imply factual data when none exists. But perhaps their point was not to either

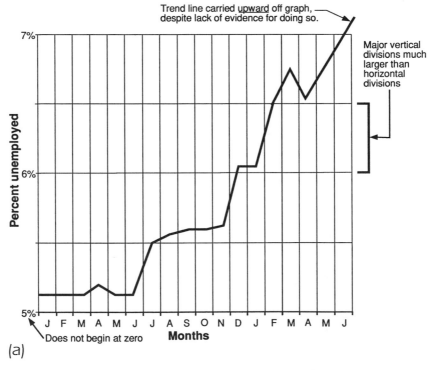

(a)

Figure 13-16 (a) How a news magazine distorts the truth. Can you count the number of graphical scams present in this graph? (b) correct plot of the data; (c) they could have made it even worse!

report the news or make predictions, but rather to distribute salt water taffy to the public.

The data are replotted in Figure 13-16b using more conventional graphing techniques: The bottom end of the scale is zero, the horizontal and vertical divisions are nearly the same size, and the trend line ends with the known data. What it shows is a *modest* increase in unemployment.

I suppose we should be happy. They could've committed a fourth fallacy. Consider Figure 13-16c, which is Figure 13-16a redrawn with a *broken horizontal scale*. Notice how the modest increase suddenly shoots up precipitously, rather than gradually increases.

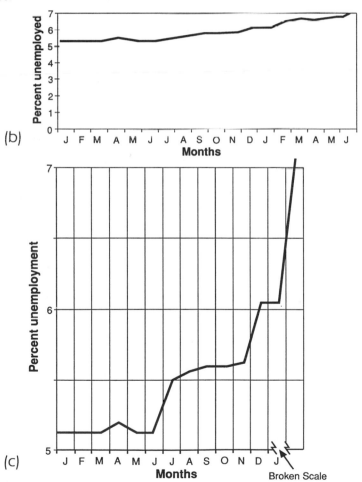

Caution: Always Graph ALL of the Data

January 28th, 1986 was a fateful day in the United States's space program. The space shuttle *Challenger*, with seven people aboard including the first "teacher in space," lifted off from the launch pad at Cape Canaveral and, little more than a minute later, exploded killing all aboard. Subsequent investigation revealed that the cause of the accident was the failure of "O"-ring seals on one of the solid-fuel booster rockets.

It is now known that the seals had a high probability of failure at cold temperatures; the temperature prior to launch was only 36 °F. What must be asked, however, is whether or not these facts were known—or should have been known—prior to the flight. According to the Presidential Commission on the Space Shuttle Challenger Accident the data were known, but misanalyzed. According to the commission's *Report*:

> The managers compared as a function of temperature the flights for which thermal distress of O-rings had been observed [author's note—see Figure 13-17a]—not the frequency of occurrence based on all flights [Figure 13-17b]. In such a comparison [Figure 13-17a], there is nothing

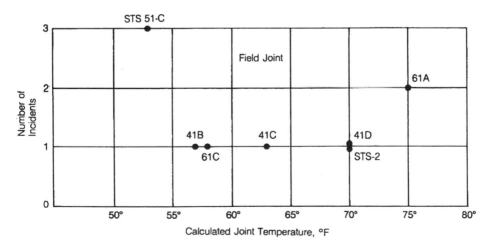

Figure 13-17 "O"-ring data from Space Shuttle Challenger: (a) Only failure data shown; (b) Both failure and non-failure data shown.

irregular in the distribution of O-ring 'distress' over the spectrum of joint temperatures at launch between 53 °F and 75 °F. When the *entire history* of flight experience is considered, including 'normal' flights with no erosion or blow-by, the comparison is substantially different [Figure 13-17b].

This comparison of flight history indicates that only three incidents of O-ring thermal distress occurred out of twenty flights with O-ring temperatures at 66 °F or above, whereas *all four flights* with O-ring temperatures at 63 °F or below experienced thermal distress.

Consideration of the *entire* launch temperature history indicates that the probability of O-ring distress is increased to almost a certainty if the temperature is less than 65 (*emphasis added*).

In other words, had all of the data been evaluated, the conclusion of managers might have been different. It would have been noted that *all* of the flights at cold temperatures—"cold" being defined as 63 °F or less—experienced O-ring failure.

It wasn't as if they weren't warned. One engineer of the rocket manufacturer is reported to have expressed concern. The

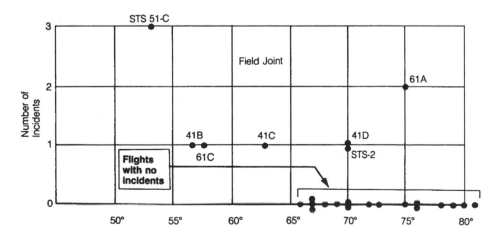

Calculated Joint Temperature, °F

atmosphere of the space program was said to have deteriorated from the original period that ended with the *Apollo* flights. In those days, the managers were often the engineers and scientists who developed the rockets... and consequently knew failure first-hand. The shuttle management was apparently self-deluded, believing beyond the reasonable data that the probability of a serious accident was a miniscule 1 in 100,000; the engineers and scientists variously predicted 1 in 55 to 1 in 70 as the range of probability of serious accident. The pressures on management were not technical, but rather they were political within the agency, and that caused them to overlook sound technical judgment.

According to the television dramatization of the accident, the engineer who warned management of the problem was told "...put on your management hat." While the statement may or may not have been actually uttered in the *Challenger* case, it is commonly heard in development projects. Such a statement to an engineer or scientist is very often an invitation to go with the tide, take the easy course, and stuff scientific integrity into a dark hole; it is the technical equivalent of "politically correct." In your professional career, if, in response to your own concerns that management considerations are taking precedence over sound scientific judgment, reply by saying: "Yes, let's launch the *Challenger* anyway."

And, whether or not the issue is as important as the lives of astronauts, when evaluating data be sure to evaluate—and graph—ALL of the data. Patterns that emerge when all data are considered may not be apparent when only selected data are presented.

Oscillographic Presentations

One of the most common ways to record data in scientific laboratories is through the use of an oscillographic recording device. Several types are used, but they break into three categories: *paper, cathode ray oscilloscope* (Figure 13-18a), and

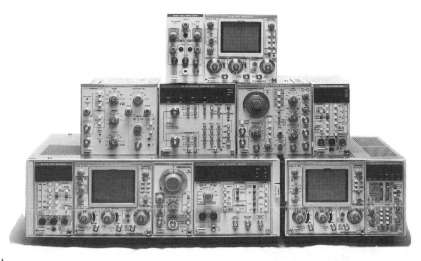

(a)

computer terminal. The paper recorders include y-time strip chart recorders (Figure 13-18b), as well as x-y and y-time fixed paper recorders. A cathode ray tube oscilloscope, or CRO, is an electronic instrument that produces either y-time or x-y recordings.

The computer form of oscillographic display breaks into both paper and CRO classes according to the type of output device used. If the computer monitor is used, then it resembles the CRO, but if a plotter or dot-matrix graphics printer is used, then the computer type is similar to the paper recorder. The instrument shown in Figure 13-18c is a medical monitor

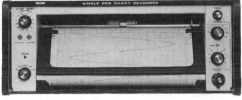

(b)

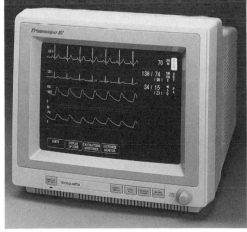

(c)

Figure 13-18   (a) Oscilloscopes; (b) X-time strip chart recorder; (c) Modern video terminal medical oscilloscope.

oscilloscope. It is computer driven, and displays the physio-logical waveforms obtained from a patient through electrodes and sensors.

Any form of oscillographic recorder will have a series of vertical and horizontal divisions that are used for measuring the units being recorded. A typical CRO, for example, has ten 1-cm divisions along the horizontal axis, and eight or ten 1-cm divi-sions along the vertical axis; minor divisions are typically five or ten per main division.

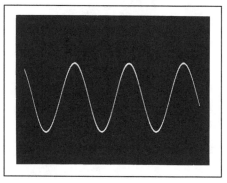

(a)

Figure 13-19 Oscilloscope presentations are actually graphs. These are Voltage-vs-time graphs of (a) a sine wave; (b) a sawtooth wave; (c) the human blood pressure wave-form taken with a photoplethysmograph on a thumb or earlobe.

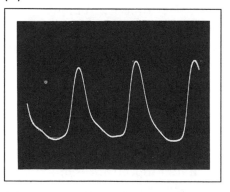

(b)

(c)

Figure 13-19 shows three different y-time traces taken from a CRO screen by a photographic camera. In Figure 13-19a the waveform is a sinusoidal waveform. By measuring along the vertical axis it is possible to measure the peak-to-peak voltage, while along the horizontal axis the period (hence also the frequency) can be measured by noting the number of divisions for each cycle, and multiplying it by the scale factor on the CRO horizontal timing controls. The waveform in Figure 13-19b is a sawtooth waveform, and it is treated similarly to the sine wave. The waveform shown in Figure 13-19c is a human peripheral blood pressure pulse taken from a noninvasive external sensor called a *photoplethysmograph* (it lacks some of the features of the arterial pulse shown earlier).

A pair of *x-y* patterns are shown in Figure 13-20. These traces are technically called *Lissajous figures*. The number of loops along horizontal and vertical edges can tell us something

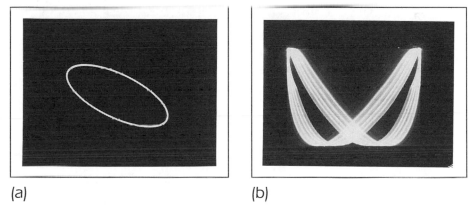

(a)   (b)

13-20   Lissajous x-vs-y oscilloscope plots. Sine wave data are presented: (a) both x and y frequencies equal; (b) One frequency maintained constant, while the other is swept back and forth across the constant frequency.

about the relationship of the frequencies applied to the *x* and *y* inputs. In Figure 13-20a the pattern is a single loop, so the frequencies applied to the two inputs are equal. But the fact that the loop has thickness (instead of being a straight line) tells us that there is a phase difference between the two signals. The "fatness" can be used to measure the phase difference. In the Lissajous pattern of Figure 13-20b the *x*-axis frequency was held constant, while the *y*-axis frequency was swept sinusoidally across the *x*-axis frequency. There are a large number of different applications for Lissajous patterns. One common use is in making *vectorcardiograms* of the heart signals from different angles with respect to the heart.

Digitizers

Computers are used extensively in science, and with the personal computer revolution every scientist can have one of these nifty machines. But an awful lot of data is taken manually, even today. In order to get that data into the computer, you can use either of several techniques. . . one of which is "fingerboning" each data point into the keyboard. Another way is to use a *digitizer*. These instruments connect to the computer and hold a graph. A pointer is placed over the data point, and a button is pressed. The digitizer tells the computer the x,y coordinates, relative to the digitizer pad, at which the data point exists. Nifty, huh?

Doing Science with Computers

NOT TOO long ago I attended a conference on radar where some very advanced signal processing papers were presented. Sitting a couple of chairs from my own was a man who is very well known to radar scientists. His books are widely read, and form the nucleus of any good professional radar library. He remarked: "You know, none of this is really new...I had my graduate students solving these problems in 1960, *but today we have the computer power to implement them without breaking the bank.*"

Indeed we do, and we are very lucky in that respect. I am typing this sentence on a desktop personal computer (PC) that has five times the speed, forty times the memory, and a zillion times the amount of available software as the first digital computer that I used in college. This little PC even has a hard disk drive, which the college computer in the late 1960s lacked altogether. Yet the PC takes only a fraction of the electrical power (the old one heated one end of the science hall), doesn't need punched cards for data input, and has worked for three years without servicing. The old computer could not make that claim. What's more, my desktop PC is "bottom of the line" and cost less than the paltry advance I received for this book.

Desktop computers revolutionized many areas of science in the 1970s, especially experimental lab science. These tools will do a lot of different things for us. They will, for example, take in the raw data and perform the tedious statistical calculations that used to take many hundreds of hours to perform. We can reduce a week's worth of tedium to less computer time than it takes to go to lunch, with smaller possibility of mistakes (notice I didn't say NO possibility of mistakes).

In addition, the computer will acquire the data for you. By installing a device called an *analog-to-digital converter* (commonly called an A/D converter), of which more later, we can change the continuous "analog" electrical signals produced by a host of different instruments and sensors into the type of data that a computer can digest. In the good ol' days, during the course of an experiment, you would periodically look at an array of instruments and write down the readings on a log sheet. The fastest sampling interval was set by how rapidly a graduate student (a lowly type of slave who sometimes pays for the privilege of laboring 12 hours a day) can jot down the numbers. And that process offered innumerable chances for mistakes! The computer equipped with an A/D converter can take data samples at intervals set by the program, down to milliseconds (10^{-3} s) or microseconds (10^{-6} s) apart.

A computer can be used in the laboratory to control an experiment or process. It can, with appropriate interfacing circuitry and software, turn things on and off, and do other chores that actually carry out an experiment. This capability not only relieves the operator of boredom (or possibly deprives a student of a part-time job), but it can also make the experiment more consistent from one trial to another. A significant source of error in scientific experiments is the artifact created when a human operator does some chore a bit differently each time it is carried out. The equivalent variation in a computer-controlled experiment is much smaller than in human-controlled cases.

A friend of mine, the late Johnnie Harper Thorne, worked for the University of Texas in Austin as an instrumentation technician, computer whiz, and all-around "catcher of bullets in his teeth" type of guy. He once integrated several instruments and machines together with a computer, so that his boss could carry out a certain experiment. The result kept coming up "6.4" on every run, while before the numbers were all over a normal distribution curve. When the scientist wondered why, Johnnie, with more than a little mischief in his heart, chuckled: "The computer prints the numbers *I* tell it to print." He explained to me that the principal source of variation in the old method was differences in the time required to manually mix two chemicals as they were injected into a container that housed a mineral or chemical sample being examined. When the computer took over, the variation reduced to such a small value that the difference were in the hundredths column, or less. When the computer was reprogrammed to print out numbers such as "6.43," then the expected variation surfaced, but the computer variation was an order of magnitude smaller than the manual method variation.

Sometimes nonvariation does not indicate a well-controlled process, but rather a broken instrument. A computerized instrument used to measure hematocrit (the percentage of blood volume that is red cells) in a medical laboratory only read out "46." Unfortunately, this value is within the normal range for both men and women, so for months no one thought to question it. That is, until a particularly bright technologist came along and realized hematocrit values should be normally distributed. Running a sample both on the machine and the old-fashioned manual way revealed a startling difference in value. He then ran a sample of tap water through the machine, and found that ordinary tap water is 46% red blood cells, if the instrument was to be believed! The problem turned out to be a defective electronic circuit in the instrument's computerized heart.

The lesson to be learned? *Don't take "Yes" for an answer when it seems contrary to what you know to be true.* Always search out the root cause of anomalies in data!

Figure 14-1 shows a hypothetical research laboratory computerized experiment based on an actual case. A desktop computer (an IBM-AT clone) was used to control an experiment in which a new drug was administered to a dog in order to discern its effects. The drug is administered through a computer-controlled pump driven by an output port of the computer. The researcher, an anesthesiologist, wanted to measure the oxygen breathed by the subject, the expired CO_2, the blood pH, blood pO_2 and some other parameters (listed here as "Blood Whatever"). All of these parameters are measured by appropriate sensors connected to instruments that process the sensor signal. The outputs of these instruments were analog voltages, so an A/D converter is needed to convert the analog voltage to a binary digital word (which can be input to the computer through an input port).

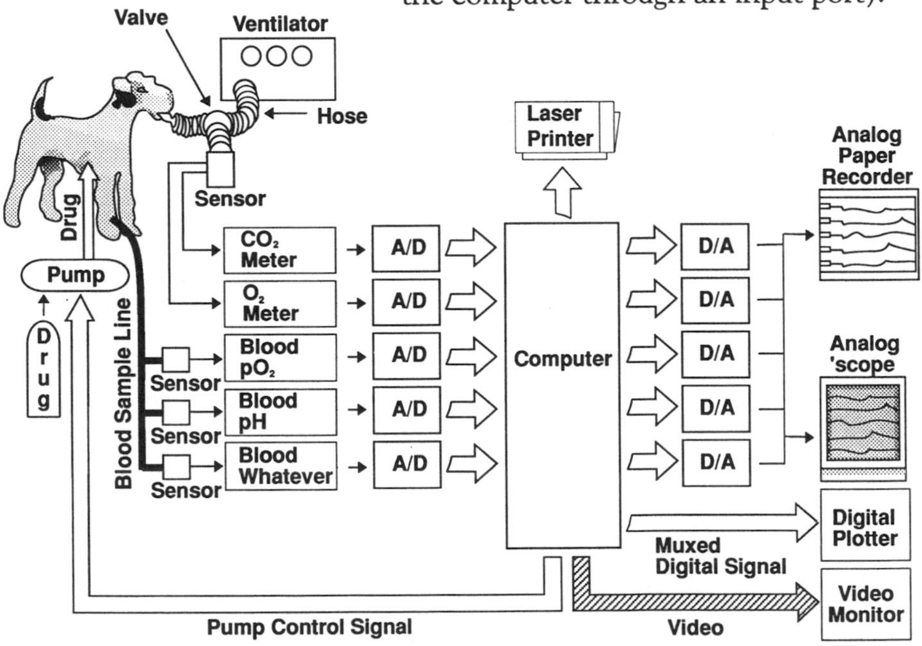

Figure 16-1   A computerized experimental laboratory.

The computer processes, stores, and does whatever is needed to put the data into the proper form. It then outputs the data to any of a variety of output devices, such as a laser or dot matrix printer, a digital plotter, or a video monitor. If analog display devices, such as meters, analog paper recorders, or an old-fashioned analog oscilloscope are used, then the computer will need a *digital-to-analog (D/A) converter* to make the conversion to an analog signal that can be used by those devices.

Depending on how the computer is programmed, it can process the data into statistical form, and produce graphs or tables to show humans what the data is doing. When it comes time to write up the results of the experiment for publication, the same computer can be used to type the paper on word processor software, prepare the camera-ready graphics on a graphic program, and then print the whole thing out "ready for publication" on the laser printer. A lot different from the bad ol' days!

One of the strongest implications of the computerization of laboratories is the possibility of doing mathematical modeling of an experiment before it is actually carried out. Experiments can be both expensive and time consuming, so doing some work on the computer before the "live" experiment saves resources and time.

You must be a bit cautious about computer modeling, however. The problem comes when researchers forget that the results printed out are a mere model, and are not reality itself. The usability of a model is heavily dependent on the assumptions made (and there are always assumptions) when writing the model, and parameters selected for input data. Real experiments can produce different results, in which case an evaluation must be done to determine which (if either) is correct. Computers are a powerful analysis and modeling tool, but do not become so enamored of glowing multicolor video screens that you forget that virtual reality is not real reality. *Science is not a video game.*

Computer Numbers: Base-2

Digital computers operate in the base-2, or *binary*, number system. While you don't have to fully understand binary numbers to gain a working knowledge of computers, you should have at least some glint of familiarity with it. We will not go into base-2 arithmetic, but must discuss the base-2 numbers simply because a computer sees the world in that frame of reference.

The binary number system recognizes only two digits: 0 and 1. In the early 1960s, mathematics professor and comedian (?!?) Tom Lehrer said that base-2 arithmetic is like the more familiar decimal arithmetic system "...if you're missing eight fingers" (perhaps he meant nine fingers). The reason for using binary arithmetic is that the computer's innards work like switches: ON (1) or OFF (0). Inside the computer, the binary digits 0 and 1 are represented by voltage levels. A common protocol is to make 0 volts represent the digit 0 and +5 volts represent the digit 1. These values are sometimes referred to as LOW and HIGH, respectively. To keep the jargon straight:

0	1
0 volts	+5 volts
LOW	HIGH
Logical–0	Logical–1

Keep in mind, however, that these values are only a convention, and that some equipment uses different values. An example is the RS-232C serial communications standard used by a lot of personal computers. It assigns binary 0 to a voltage between +3 and +12 volts, and binary 1 to a voltage between −3 and −12 volts. As long as a system is self-consistent, or uses some sort of translation between protocols, then it will work well. The only time trouble usually occurs is when you attempt to interface two equipments with differing binary/logic definitions. Be especially wary of old clunkers hauled out from under a dumpster out back of the college science building, especially if they were made prior to 1970.

In a computer context, a base-2 digit is called a *bit* after the name *binary digit*.

In the same manner as for decimal numbers, base-2 numbers in combination with one another are given added weight. The decimal digits are 0, 1, 2, 3, 4, 5, 6, 7, 8, and 9. But when two decimal numbers are placed side-by-side, they gain added value. For example, the decimal number "62" takes on the value of $(6 \times 10^1) + (2 \times 10^0)$. The same thing happens for binary numbers, in which the position of the digits (0 and 1) add value. For example, the binary number "101" means $(1 \times 2^2) + (0 \times 2^1) + (1 \times 2^0)$, or $(4 + 1) = 5$. Thus, the binary number "101" is the same value as decimal "5."

In computer literature it is the common practice to array bits into *words* of a standard bit-length. For example, the early microprocessors used an 8-bit word ranging in value from 00000000 to 11111111 (0 to 256 decimal). An 8-bit array of bits is called a *byte* (although unofficial, some people call a 4-bit array a *nibble*). Later computers use 16-bit words:

$$1111111111111111$$

or 32-bit words:

$$11111111111111111111111111111111$$

In dealing with computers you will see the terms *most significant bit* (MSB) and *least significant bit* (LSB). The MSB is the highest-value bit in the binary number, i.e., the leftmost, while the LSB is the lowest-value bit (rightmost).

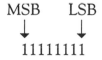

In the case of an A/D converter, used to input electrical signals to the computer, there is a minimum step voltage that can be discerned. This voltage, called the *least significant bit voltage* (1-LSB voltage), establishes the measurement resolution of the system.

In other cases, the computer literature specifies the *octal* (base-8) and *hexadecimal* (base-16, commonly called "hex") number systems. The octal digits are 0 through 7, while the hex digits are 0 through 15 with 10 through 15 being represented by the letters A–F (i.e., 0, 1, 2, 3, 4, 5, 6, 7, 8, 9, A, B, C, D, E, and F). The internal workings of the computer are still in binary, but for convenience the documentation of the computer may list binary values in their octal or hex (of which hex is now the most common) equivalent. For example, the maximum value of an 8-bit word (11111111) could be represented as "FF" in hexadecimal.

In some texts a notational convention is adopted to distinguish the number systems being used in order to avoid confusion. For example, "101" could be "one hundred and one," or it could be the binary equivalent of decimal "five," or the octal equivalent of decimal "one hundred forty-five." In order to avoid confusion, some texts use a subscript to denote the base. For example, "101_{10}" is decimal, "101_2" is binary, "101_8" is octal, and "101_{16}" is hexadecimal. In some cases, hexadecimal numbers are denoted with a prefix or suffix such as "101H" or "101$" or "H101."

Analog-to-Digital Converters

As mentioned, the analog-to-digital converter (A/D) is an electronic circuit that will receive a continuous analog input voltage (V_{in} in Figure 14-2), and convert it to an n-bit binary (base-2) digital output. This output signal can be applied to the input port of a digital computer.

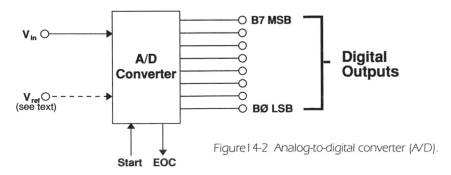

Figure 14-2 Analog-to-digital converter (A/D).

It is important for scientists and engineers to understand a bit about how A/D converters work, since they determine exactly what numbers are assigned to your data when it enters the computer. A typical A/D converter uses a START signal to initiate conversion, and then issues an EOC (*end-of-conversion*) signal when the conversion is completed. The data on the A/D output lines is not valid until EOC is issued. The time period between the START signal and the EOC signal is the *conversion time* (t_c) of the A/D converter. Typical conversion times range from one microsecond (1 µs) to several dozen milliseconds (ms), although expensive nanosecond-level converters can be purchased.

A/D Digital Coding Schemes

There are several different coding methods for defining the binary number output from an A/D converter. The coding scheme is the protocol that determines which binary codes represent specific voltage levels within a specified range. The most common of these are: *unipolar positive, unipolar negative, symmetrical bipolar*, and *asymmetrical bipolar*. In each case, the idea is to represent the range of voltages or currents using the maximum number of bits available in the computer.

The unipolar coding schemes provide an output voltage of one polarity only. In these circuits 0 volts is the minimum and some positive or negative value is the maximum. Because one binary number state represents 0 VDC, there are $[2^N - 1]$ states to represent the analog voltages within the range. For example, in an 8-bit system there are 256 states, so if one state (00000000_2) represents 0 VDC then there are 255 possible states for the non-zero voltages. Thus, the maximum output voltage is always 1 LSB less than the reference voltage. For example, in a 10.00 VDC system, the maximum output voltage is 10.00 volts/256 = 0.039 volts = 39 mV less than the reference voltage. Thus, the maximum output voltage will be approximately 9.96 VDC. The unipolar positive coding scheme is shown in Table 14-1.

Unipolar Output Voltage	Binary Input Word
0.00 VDC	00000000_2
$\frac{1}{2}V_{max}$	10000000_2
V_{max}	11111111_2

Table 14-1

The negative version of this coding scheme (unipolar negative) is identical, except that the midscale voltage is $-\frac{1}{2}V_{max}$ and the full-scale output voltage is $-V_{max}$. A variant on the theme inverts the definition, as seen in Table 14-2.

Unipolar Output Voltage	Binary Input Word
0.00 VDC	11111111_2
$\frac{1}{2}V_{max}$	10000000_2
V_{max}	00000000_2

Table 14-2

The bipolar coding scheme faces a difficulty that requires a trade-off in the design. There are an even number of output states in a binary system. For example, in the standard 8-bit system there are 256 different output states. If one state is selected to represent 0 VDC, then there are 255 states left to represent the voltage range, which is an odd number. As a result, there are an odd number of states to represent positive and negative states either side of 0 VDC. For example, 127 states might be assigned to represent negative voltages, and 128 to represent positive voltages. In the asymmetrical bipolar coding, therefore, the pattern might look like that shown in Table 14-3.

Unipolar Output Voltage	Binary Input Word
$-V_{max}$	00000000_2
0.00 VDC	10000000_2
$+V_{max}$	11111111_2

Table 14-3

A decision must be made regarding which polarity will lose a small amount of dynamic range. The other bipolar coding system is the symmetrical bipolar scheme. The decision in the symmetrical scheme is that each polarity will be represented by the same number of binary states either side of 0 VDC. But this scheme does not permit a dedicated state for zero. The scheme is shown in Table 14-4.

Unipolar Output Voltage	Binary Input Word
$-V_{max}$	00000000_2
$-$Zero (-1 LSB)	01111111_2
0.00 VDC	(disallowed)
$+$Zero ($+1$ LSB)	10000000_2
$+V_{max}$	11111111_2

Table 14-4

The state "plus zero" is more positive by the 1-LSB value than 0 VDC, while the "minus zero" state is more negative than 0 VDC by the same 1-LSB value.

A/D Converter Resolution

The resolution of the A/D converter was discussed above in one sense—namely, the smallest increment of signal voltage that can be discerned is set by dividing the analog signal voltage range into $N-1$ 1-LSB steps. Table 14-5 shows the resolution capability of A/D converters from 0 to 16 bits word length. The 2^N column shows the number of different levels that are available (the number of intervals is one less than these numbers). The value of each interval is given by the 2^{-N} column. To find the 1-LSB voltage represented by each interval multiply the voltage range by the number found in the 2^{-N} column for the given bit-length (N).

Table 14-5

Bits	2^N	2^{-N}	dB	Percent	PPM
0	1	1.00	0	100	1,000,000
1	2	0.50	−6	50	500,000
2	4	0.25	−12	25	250,000
3	8	0.125	−18.1	12.5	125,000
4	16	0.0625	−24.1	6.2	62,500
5	32	0.03125	−30.1	3.1	31,250
6	64	0.015625	−36.1	1.6	15,625
7	128	0.007812	−42.1	0.8	7,812
8	256	0.003906	−48.2	0.4	3,906
9	512	0.001953	−54.2	0.2	1,953
10	1,024	0.0009766	−60.2	0.1	977
11	2,048	0.00048828	−66.2	0.05	488
12	4,096	0.00024414	−72.2	0.024	244
13	8,192	0.00012207	−78.3	0.012	122
14	16,384	0.00006104	−84.3	0.006	61
15	32,768	0.00003052	−90.3	0.003	31
16	65,536	0.00001525	−96.3	0.0015	15

The *decibel (dB)* column represents the 2^{-N} data to log form, which is used in some signal processing and instrumentation applications. The decibel value is found from:

$$dB = 20 \log (2^{-N} \text{value})$$

In some cases, the resolution is expressed in terms of a percent, which is found from:

$$r = \frac{100\%}{2^N} = (100\%)(2^{-N})$$

In most A/D circuits there is an analog reference voltage (V_{ref} in Figure 14-2) that is used to compare the unknown analog input signal voltage, although in many integrated circuit A/D converters the reference voltage is internal. No matter how many bits the A/D converter claims, the maximum resolution is limited to the accuracy of the reference voltage. If this voltage drifts with time or temperature (often the case), or if it is in error to start, then the resolution of the A/D converter is limited. For example, in an 8-bit A/D converter used to measure a 0–10 volt range, the 1-LSB value is 39 mV. If the reference voltage varies by, say, 60 mV, then the device can only provide 7 bits of resolution reliably. In some A/D converter specification sheets you will see the device listed as "10-bits resolution with internal reference, up to 12-bits with suitable external reference source." This indicates the limitations of an internal reference source, and even though there may be 12 data lines coming out of the device, its true resolution is only 10 if that source is used.

The resolution expression is altered to account for the uncertainty in the reference voltage:

$$r = [(2^{-N} + \Delta V_{ref})(100\%)]$$

Quantizing Error

Because there is an inherent minimum voltage step (the 1-LSB value) associated with A/D converters, there is also a *quantizing error* inherent in the process. Notice in Figure 14-3 that the analog signal voltage can take on any value between 0 and 7 volts. But the 3-bit A/D converter (acting as a "quantizer," in essence) can only accommodate the discrete values 0, 1, 2, 3, 4, 5, 6, or 7 volts. If the input voltage is, say, 2.56 volts, then the input voltage lies between two of the permissible output values (2 and 3 volts). In this particular example, the output from the A/D converter will be 101_2, indicating 3_{10}. There is a decision level value midway between the allowable values where the A/D output will snap to the next higher or lower value, depending on which side of the decision level the voltage is on. If we denote the quantizing error by Q, then the error will be $\pm 1/2 Q$ at worst case, falling to zero only at those points where the analog input voltage exactly corresponds to the allowable binary value.

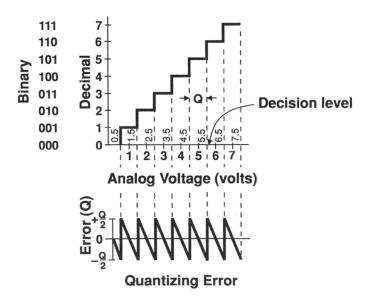

Figure 14-3 Quantization error is created because the digital outputs of the A/D converter can only take on discrete values.

Aperture Time Error

Another source of inherent error in real data converters is caused by the fact that the conversion process takes time. This time period is called the *aperture time* (t_a). The most obvious portion of the aperture time results from the actual conversion time (t_c), but there are also other contributors that may affect t_a in some cases. In Figure 14-4 we see how the aperture time error affects the result. Signal V_{in} is applied to an A/D converter that takes t_a to do its job. In the meantime, however, the signal voltage changes an amount ΔV... which represents a maximum error value. The A/D output is ambiguous by the factor ΔV.

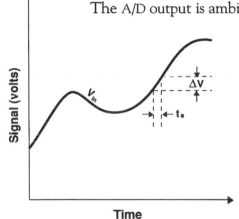

Figure 14-4 Slope of a curve at a point is the change of amplitude over the change of time ($\Delta V/\Delta t$). The value of t_a must be kept small enough so that A/D conversion doesn't result in a large error because the ΔV term is large.

Last Digit "Bobble" Error

Because of aperture-time errors and quantizing errors, not to mention spurious noise on the input signal, the last digit of a binary word produced by an A/D converter is always regarded as ambiguous. Depending on how close the actual voltage plus noise is to the decision point, or how much it changes during the conversion period, or how the noise affects its actual algebraic value, the LSB might flip one way or the other. Thus, a value of "11111111_2" (256_{10}) could actually be "11111110_2" (255_{10}). This phenomenon is called *bobble* because in real converters the LSB tends to flip at random between 0 and 1.

Last-digit bobble must be considered when determining the resolution required for any given application. If you need 8 bits of resolution, then (without special encoding, compression, or correction schemes) plan for 9 bits or better.

Sampled Signals

The digital computer cannot accept analog input signals, but rather requires a digitized representation of that signal provided by the A/D converter. If the A/D converter is either clocked, or allowed to run asynchronously according to its own clock, then it will take a continuous string of samples of the signal as a function of time. When combined, these signals represent the original analog signal in binary form.

But the sampled signal is not exactly the same as the original signal, and some effort must be expended to make sure that the representation is as good as possible. Consider Figure 14-5. The waveform in Figure 14-5a is a continuous voltage function of time, $V(t)$; in this case a triangle waveform is seen. If the signal is sampled by another signal, $P(t)$, with frequency F_s and sampling period $T = 1/F_s$, as shown in Figure 14-5b, and then later reconstructed, the waveform may look something like Figure 14-5c. While this may be sufficiently representative of the waveform for many purposes, it would be reconstructed with greater fidelity if the sampling frequency (F_s) is increased.

Figure 14-6 shows another case in which an analog signal, $V(t)$ in Figure 14-6a, is sampled

(a)
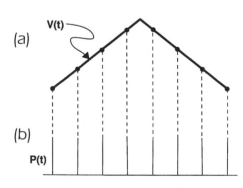

(b)

(c)
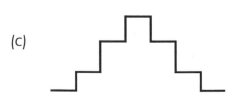

Figure 14-5 (a) Triangle waveform; (b) sampling pulses; (c) reconstructed waveform.

by a pulse signal, $P(t)$ in Figure 14-6b. The sampling signal, $P(t)$, consists of a train of narrow pulses equally spaced in time by T. The sampling frequency F_s equals $1/T$. The resultant is shown in Figure 14-6c; the amplitudes of the pulses represent a sampled version of the original signal. The higher the sample rate, the more faithful the sampled representation to the original signal.

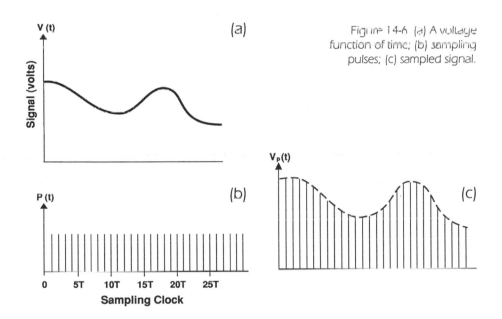

Figure 14-6 (a) A voltage function of time; (b) sampling pulses; (c) sampled signal.

The sampling rate, F_s, must (by *Nyquist's theorem*) be at least twice the maximum frequency (F_m) contained in the applied analog signal, $V(t)$.

When the sampling frequency F_s is less than twice the maximum frequency F_m, a phenomenon called *aliasing* can occur. That is, when the sampled signal is recovered by low-pass filtering it will produce not the original sinewave frequency F_o but a lower frequency equal to $(F_s - F_o)$...and the information carried in the waveform is lost or distorted.

The solution, for high-fidelity sampling of the analog waveform for input to a computer, is to:

1. Bandwidth-limit the signal at the input of the sampler or A/D converter with a low-pass filter with a cut-off frequency F_c selected to pass only the maximum frequency in the waveform (F_m) and not the sampling frequency (F_s). (For those not electronically inclined, consult an electronics technician or a reference on signal filtering.)

2. Set the sampling frequency F_s *at least* twice the maximum frequency seen in the applied waveform, i.e., $F_s \geq 2F_m$.

Note: Experience has shown that some users will not accept a reconstructed sampled waveform if the sample rate is limited to $2F_m$. For example, medical electrocardiograph (ECG) waveforms—in which F_m is 100 Hz—tend to look "blocky" (as one nurse remarked) when sampled at 200 Hz and then reconstructed. The user acceptance was much better when the waveform was sampled at 500 Hz, or $5F_m$. While that rate was expensive to accommodate in an 8-bit A/D converter at one time, it now costs very little and should be used. . . Nyquist's criterion notwithstanding (this criterion refers only to technically competent data sampling, not to user acceptance of the reconstructed waveform).

Multiplexed A/D Converters

In the hypothetical computerized laboratory shown earlier there was a dedicated A/D converter for each channel of analog signal. That is often the best way to design a system, especially now that common 8-bit A/D converters are relatively cheap. In some cases, however, it may be better (or more economical) to use a *multiplexer* to sequentially connect all analog input signals to a single A/D converter. Figure 14-7 shows that the multiplexer (commonly called "MUX") is a single-pole, multiple-position switch. A command signal from the controller circuit sequentially steps the MUX switch through all positions (eight different positions in this example). Up to eight different instruments can be connected to the same A/D converter.

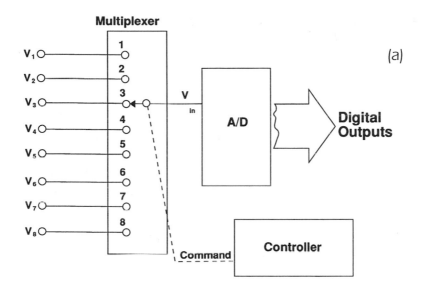

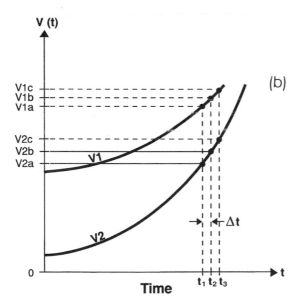

Figure 14-7 (a) Using a multiplexing switch to make a single A/D converter work on multiple input voltages; (b) potential error source.

The MUXed A/D converter is a wonderful little tool, but there is a potential problem with fast-changing signals. Consider Figure 14-7b in which two related signals (V_1 and V_2) are MUXed to the same A/D converter. Suppose the experimenter wants to know the exact values of V_1 and V_2 at the same time. Unfortunately, the A/D converter can only sample one signal at a time (remember, MUXed signals are examined sequentially), and there is a conversion time Δt associated with how long it takes the A/D to make a conversion. If V_1 is sampled at t_1, then the sample of V_2 is taken at $t_1 + \Delta t$, or t_2. The value of V_1 at t_1 is V_{1a}, which should be associated with V_{2a}. At t_2, however, V_2 has changed value to V_{2b}. That creates an error in the values that will be associated together when the data is analyzed. If Δt is sufficiently small, compared with the rates of change of V_1 and V_2, then there is no problem—the error is negligible. But if Δt is long compared with the rates of change of V_1 and V_2, then a significant error may result.

A possible solution used in many computers is shown in Figure 14-8a. In this two-channel MUX system, V_1 and V_2 are fed to an analog *sample-and-hold* (S&H) circuit (SH1 and SH2). These circuits take quick "snapshot" samples of the analog signals simultaneously. These signals are held until the MUX and A/D have time to cycle through both inputs.

A graphical illustration of how an S&H circuit works is shown in Figure 14-8b. The dotted line represents the analog input signal, V_{in}, while the solid line represents the output voltage from the S&H (V_o)—which is what the A/D converter sees. The S&H sampling signal is an electrical pulse that commands the S&H circuit to take a sample. Voltage V_o rises from zero to the level of the first sample between time t_1 and t_2. During the time that the command is in the SAMPLE state, the output will track the input signal, but when the command goes back to the HOLD state, it will remain at the last value of the input signal. The output signal does not change again until a new sample pulse period arrives (t_3–t_4 or t_5–t_6).

Figure 14-18c shows the signal discussed earlier in Figure 14-6, but with a "sampled and held" protocol rather than an instantaneous sampling. This version is more like what is seen in real situations. Notice that an error exists because the S&H interval duration is too long.

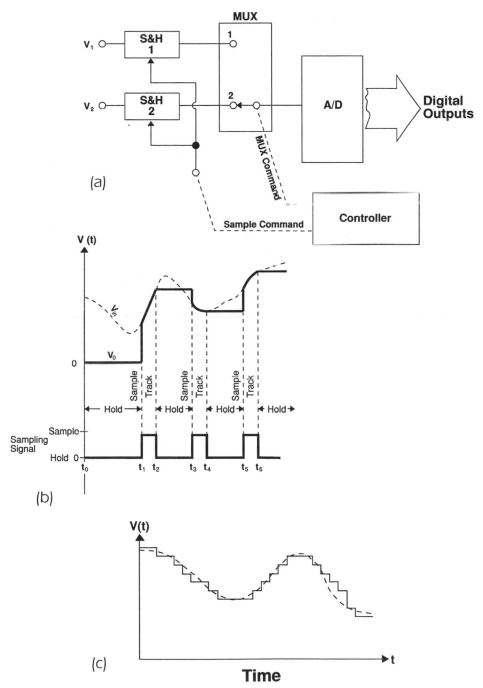

Figure 14-8   (a) Sample and hold (S&H) system to reduce error term; (b) action of the S&H circuit; (c) Output of an S&H as the function (dotted line) varies.

Other Error Sources

One of the errors that crops up in computerized instrumentation is a special sampling error caused by the sampling rate and the sampling synchronization in scanned measurements. Let's take a look at a practical example.

Consider a scanning optical densitometer such as Figure 14-9. In this instrument a high-precision optical sensor scans across a piece of exposed and developed photographic film. The output of the photodetector is a continuous analog electrical signal that represents the amount of light passed through the film. Higher density sections of the film transmit less light than lower density sections, so we are able to plot a profile of the exposed film density as the film is translated (Δx) beneath the photodetector.

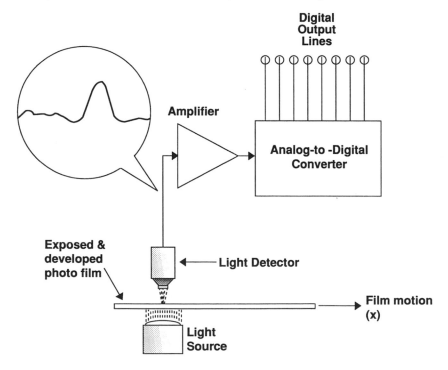

Figure 14-9 Optical densitometer instrument.

The *resolution* of the measurement is the ability to separate two targets Δx apart (Figure 14-10). If we claim, say, a 1-mm

resolution, then two equally dense spots on the film 1 mm apart
will be discerned as two separate targets by the system. If the
resolution is actually worse than 1 mm, then two targets merge
into one in the
computer data.

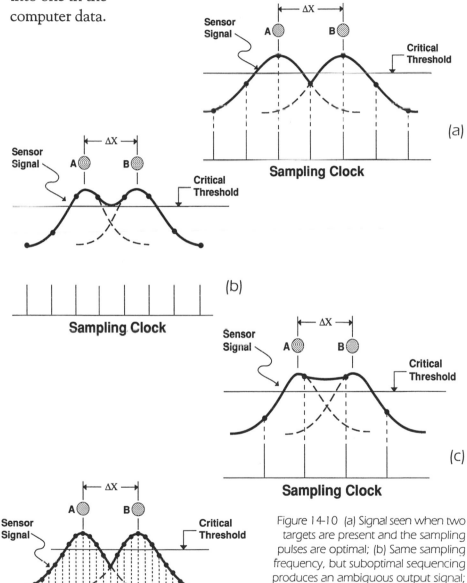

Figure 14-10 (a) Signal seen when two
targets are present and the sampling
pulses are optimal; (b) Same sampling
frequency, but suboptimal sequencing
produces an ambiguous output signal;
(c) too few sampling pulses really mess
up the output; (d) A high sampling
frequency increases the probability
of seeing both targets.

In most computer-based instruments, a threshold value change must be crossed in order to see a difference on a display device (such as a video screen). If the signal does not drop below that value as the detector moves between targets, then the operator will see only one target. In some systems, the critical threshold is a 2:1 change in brightness, or a change of some predetermined number of gray shades. Or it might even be one picture element ("pixel"), which is the best that it can get.

Digitized data is always sampled, either through the action of a sample-and-hold circuit, or by virtue of the fact that each A/D conversion takes a finite period of time. If the sampling frequency is high enough and the samples occur at exactly the right time as the detector sweeps (as in Figure 14-10a), then the resolution criterion is met. But look what happens in Figure 14-10b at the same sampling frequency, but with the synchronization off. The depth of the notch and the actual centroid of each response broadens to a point where the operator may see one target, but may also see two targets. In Figure 14-10 neither the sampling rate nor the synchronization are up to the chore, so the two targets clearly merge into one.

The solution to this type of error is to sample at a high rate so that we are ensured of a sample occurring very near the notch and at a large number of points on each curve (Figure 14-10d). The samples might miss the actual notch at its deepest point, but there will be at least one or two samples taken below the critical threshold.

How to Win a Science Fair—
A Guide for Younger Readers

SCIENCE FAIRS are a popular annual event for aspiring amateur scientists—many of whom are future professional scientists—in grades 7 through 12. When I was in junior and senior high school, I entered the science fairs every year. For the past several years I've been privileged to serve as a judge in a local middle school and area-wide science fairs—and my interest seems more intense today than when I was a student participant. It is a real thrill to see youngsters chipping away at the edges of science, venting some of their inherent curiosity, and possibly preparing themselves for a career in science, medicine, or engineering.

The science fair sequence in most localities starts at the local school level with a school-wide science fair where outside judges and possibly some teachers select first-place, second-place, third-place and honorable-mention awardees. The first-place awardees are sent forward to an area-wide science fair where they compete against students from other schools in the locality. The winners of the area events meet in state-wide or regional science fairs, and the winners of those fairs go on to a national event.

The science fair must, unfortunately, have both winners and losers. Of course, we don't call anyone a "loser." We give all of them a certificate of attendance to reflect their hard work. The purpose of the fair is educational, after all. But it's still

nice to take home at least an honorable mention award—and
of course a first place award is golden!

Some Do's and Don'ts

Over the past several years I've noticed a few patterns
among the winners, and these observations may help *you* win
your next science fair. Let me tell you about two different
students. I'll call them John and Joan (not their real names),
and I'll disguise their projects enough to preserve their privacy.

John is a bright 12th grader who (he told us proudly) had
taken all four of the science courses offered in his school: biology,
chemistry, physics, and advanced biology. He built a project
that demonstrated a well-known scientific principle—it basically
repeated in more sophisticated form one of the experiments performed in high-school physics classes. John's display included a
computer with color monitor (although, the computer turned
out not to be needed), and a very costly (about $4,000) cathode
ray tube oscilloscope. John's display poster panels were well
made using the same kind of styrofoam-backed poster board
that's used in professional advertising displays. The panels of the
posters were produced on an Apple Macintosh computer, and
printed out on a LaserWriter printer. They really looked sharp.

Joan's entry was a bit different. She had investigated a new
science called "fractals" in an experiment (not a computer
simulation) that showed the irregular patterns of the fractals,
and demonstrated how changing certain well-defined parameters of the experiment resulted in wildly different patterns.
Joan's apparatus was homemade, and was held together with
black electrical tape. It looked, I'm afraid, a little bit sloppy.
Her posters were made of ordinary poster paper that had to be
propped up with flat wooden sticks. The panels were typed on
a moderately good, but old-fashioned, typewriter that obviously needed a little mechanical attention...and two of the
eight sheets were handwritten. She had the laboratory note-

book that showed her earlier work, and all of the false starts (several of which were abject failures).

At the morning judges' committee meeting, we found that we could send any number of students ahead to the afternoon judging (not true every year). That second judging (in the afternoon) would determine the winners who would carry the local banner forward to the state or regional science fair. (In past years, the rules limited us to one entry per scientific category, but this year we had no limit on how many could go forward from the morning session to the afternoon session.) After all of the thinking and talking was over, we sent two students ahead to the afternoon judging. Joan was one of them, but John didn't even place. He left the science fair with only a certificate of attendance. Despite his computer-generated graphics, and high-priced electronic equipment, he didn't even get an honorable mention award. *Why?*

The judges' comments were illuminating. It was noted that John's computer was being used as little more than a high-priced audio oscillator, and that the oscilloscope could easily be replaced with a simple AC voltmeter. Both instruments, it was felt, were intended to impress the judges, and were not necessary to the experiment. We weren't impressed. The glitz would not hurt John's case, however. What did him in was deeper than surface glitter.

Furthermore, John didn't know how the computer generated the audio tones, and he was all but totally in the dark on how to operate the oscilloscope. He showed real distress when one judge (me) tried to adjust the horizontal time base controls on the 'scope. He seemed painfully aware that he lacked the technical expertise to reset the controls if I didn't return them to the right settings. Someone apparently had set them for him.

It was also noted that John had an attitude problem. His cocky arrogance made him seem totally disdainful of the science fair process in general and the judges in particular. One judge (not me) reported that John made a remark about his project being a "shoo-in" for first or at least second place. Not so...he lost utterly.

Originality Counts...

But all that aside, John could have still won at least a third place or honorable mention award—even with his attitude problem—if he hadn't made some really terrible mistakes. First, his project was a mere copy of an ordinary classroom demonstration. That is a bad approach for any student, although it is not always fatal. If you ape some other experiment, then at least extend it a little bit, do it a little better than usual, or somehow make your contribution more than merely following a teacher's lab notes. Only in the lowest grades, where students have only limited exposure to science, is copying an experiment reasonable. If John had been an 8th-grader, rather than a high-school senior, then we might have been impressed.

Understand Your Experiment

Second, John was *unable to explain just what the experiment was doing!* He had at least some faint idea, but not at a level that one would expect from a high-school senior. Furthermore, he failed to take into account, or at least mention, some easily observed critical elements of the experiment that would explain why his results departed from the theoretically predicted results. He would have had a shot at a second-place award if he had known about sound-wave scattering and the effects of non-sinusoidal waveforms on the experiment.

The discussion of Joan's work was a little different. Although the judges privately criticized her apparently sloppy display presentation, they also waved it aside in favor of the positive aspects of the experiment. It was noted, for example, that she had read the two basic books on chaos theory and fractal geometry that are easily available in local libraries. While it would have been better if she had also read some of the more advanced texts, no one thought ill of her for the failure. Those books are hard to find in local libraries, expensive to purchase ($75 for one!), and were probably mathematically beyond all but the very brightest—and rarest—of high-school students. It was also noted that her written report mentioned that she had sought the advice of a local mathematician who taught a fractals course in a university.

Joan had created a clever experiment that investigated a complex phenomena. While the project would not win a Nobel Prize, it was within the grasp of a high schooler, and yet was nonetheless clever and innovative. She had *asked a question of nature*, and then created an experiment that would answer that question "yes" or "no."

Do Your Own Work

Another plus in Joan's favor, noted by several judges, was that she had failed in a number of attempts at making the experiment. However, all of her failures were documented in her notebook, proving that the final result—clever as it was—was her own work. It also showed tenacity, and the fact that she was interested in the experiment, not simply in winning the prize. You see, there are no failures in science. There are failed experiments, to be sure, but each of these tells us something about nature...
even if only, "*That* wasn't the right approach!"

Joan was also able to answer more than the most basic questions about fractals.

She could either explain some anomalies seen in the experiment, or describe a method for doing an experiment to examine the issue. Some of the judges had only heard a little about this new field, so they asked her some probing questions. Her prompt answers made sense to the judges. One judge, who knew more about fractals than Joan, asked some even deeper questions. To one difficult question she gave a perfectly valid answer: *"I don't know"*... but, she averred, "I'm going to look into that next week... it sounds interesting." Don't try to snow the judges with a guessed answer... it won't work. Any qualified judge knows that the very root of scientific investigation is those three little pregnant words, *I don't know*: the Confession of Ignorance. It is the energy and curiosity to resolve that ignorance that science-fair winners and Nobel Laureates are made of. The integrity to confess ignorance, plus a bit of hard work, leads one to profound truths.

Ask a Question of Nature

Good science research calls for the experimenter to formulate a hypothesis (which means "ask a question of nature"), and then create an experiment that will yield one of two answers: Yes or no (or true/false). The question should be narrowly and clearly formulated. "Global" questions tend to be difficult to pin down in an experiment, so are likely to fail in the end.

It is not necessary to make the question asked of nature profoundly deep, for no one expects high schoolers to break totally new ground (it could happen, however!). For example, you could ask nature whether or not intense ultraviolet light affects the growth of poison-ivy plants. Phrase the question something like this: "Intense ultraviolet light [*slows/speeds up*] the growth of poison ivy plants: True or false". Or, alternatively, use the null hypothesis: "Intense UV light *does not* [*slow/speed up*]...."

Expect to address some issues in designing the experiment. For example, what wavelength ultraviolet radiation is used? Is it filtered daylight, UV A, UV B, is it the spectrum generated by a certain brand of UV lamp? Second, ask yourself how the intensity of the UV light is to be maintained constant, and/or measured. With a modified photographic light meter? A special UV meter? With a photographic light meter equipped with a filter that passes UV but screens out other light? How much time each day are the plants exposed to UV light? In how many exposure sessions, of how long each?

Be sure to include at least one control group. A control group is one that is subjected to all conditions that the test group is subjected to, *except* the condition being examined (for example, in our hypothetical situation, UV light). Make all of the same observations on both control and test groups.

Sometimes, two control groups are used. When I worked in a medical center, I noticed that our physician researchers often used a test group and two control groups to test new treatments. One group received the new drug, one group received the normal "drug of choice," while the third group received a placebo. The results of all three groups were compared.

There is also a "double blind" protocol in science that is very desirable. In order to prevent the person who measures the progress from allowing his or her own preconceived opinions to affect the outcome, no one who either administers the drugs or evaluates the patient responses is allowed to know just who was in which group. Double blind protocols cannot always be followed in a science-fair level experiment, but it is desirable if possible.

The idea in a scientific experiment is to keep all factors constant, except the one under test. For example, in the poison-ivy experiment, the UV light is the variable. The experimenter has to be certain that the temperatures of the two plant beds are the same, that the watering and feeding are the same, and that all factors, other than the amount and wavelength of light impinging on the test plants, are the same for both populations.

Keep a Notebook

Keep a bound laboratory notebook to document your work. If you want to be really classy, then go to a professional engineering supplies stores, university book store, or drafting supplies outlet and buy a real scientist's lab notebook. If that is not to your liking, or too expensive, then use an ordinary bound composition notebook from the school supplies section of the local drug store. The idea of using a "bound" book is to be able to demonstrate that no embarrassing results were removed from the notebook...a good scientist documents the failures as well as the successes.

Write out the question that you are asking of nature on the first page of the notebook. Next, write a description of the experiment (include any drawings or charts needed). Describe a protocol of how the experiment is to proceed. It is important to remain consistent throughout the experiment regarding data collection, maintenance of the experiment (for example, feeding or watering the poison ivy), and in some cases the timing of observations (how do you know how many millimeters of growth are seen per week if you don't measure the growth at the same time every week?).

As questions occur to you, enter them in the notebook and try to make some effort to find the answers . . . but don't interrupt the flow of the experiment to answer new questions. It is axiomatic in science (where real labs depend on grant money to survive) that the experiment should raise as many questions for further research as it answers...otherwise you are out of business next year.

When preparing your report and display, keep in mind that judges are not *overly* impressed by well-done graphics...but they help. My colleagues criticized Joan's display for its sloppiness. One comment was that "perhaps she was as sloppy in some of her observations." It is unlikely that the really good project will be denied a prize because of a little sloppiness on the graphics, but it is always good to put your best foot forward. In those years when the judges are limited in the number of entries they can recommend for the second judging, those little extra touches might make the difference between two students with roughly equal projects. The main idea here is to not give the judges any reason to shave a point or two off your score. While the scoring tends to be *ad hoc*, it's nonetheless real... and a sloppy poster or illegible report costs points.

Perhaps most important is that the display, the report, and your talk with the judges should be conservative in the claims that are made. Don't make any claim for your experiment that cannot be supported with data from your laboratory notebook, other pertinent records, or from "the literature" on the subject. Results are valid only when documented!

Statistics Are Important

Make sure that the number of data points taken in the experiment are sufficient to make the result statistically significant. For example, don't expect to find the effects of UV light on poison ivy plants when the control group and test group consist of one plant each. Make the group as large as possible. One of the main mistakes seen in experiments is generalizing to the whole population what only applies to a few individuals. The way to guard against this problem is to examine many individuals to ensure that the sample is typical of the "universe" of individuals.

Also, please, please, please, if your experiment results in numerical data, or requires calculations, then be sure to understand the concept of *significant figures*. A calculator with a 12-digit display does not mean that either accuracy or precision results from actually displaying all of those digits. One of John's mistakes was that he listed a value as "13.4344359635" when "13.4" was the limit of his experimental apparatus.

If your experiment results in statistical data, then learn a little bit about statistics. Don't depend solely on mean or median calculations unless you have good reason. Look into the relevant chapters of this book to learn a bit about standard deviation, variance, and other factors that professional scientists use in analyzing data. Almost any local library will have elementary statistics textbooks, and most high-school science teachers know more than enough to advise you in these matters.

The Judges Are Your Friends!

Finally, on the BIG DAY when you actually set up the display and confront—*groan, shiver*—the JUDGES, try to remain calm... and get some sleep the night before. The judges are not ogres, and they are not going to bite you (even a little bit). Every judge that I've met over the past several years has been interested in seeing youngsters succeed in science...for youngsters are the future of the profession. They may make suggestions for your future research or study. Don't be afraid to ask *them* questions! Science is as much an exchange of ideas between people as it is experiments in a solitary laboratory...and the process can be invigorating.

Conclusion

The successful science fair experiment will be well focused, properly carried out, documented in painstaking detail, and properly described in the attached report. You will find that the judges are favorably impressed by good scientific procedure, and are not impressed at all by a project that is well presented—but either improperly executed or too trivial. A simple, but properly executed, grade- or age-appropriate project can be a winner if you do it right. *Good luck.*

APPENDIX A

Counterfeits of Truth
Recognizing and Dealing with
Logical Fallacies in Science

Fallacy

*A deceptive appearance; a false idea, a mistaken belief; of
erroneous character; an argument that fails to satisfy the
conditions of valid or correct inference, i.e. an argument in
which all premises are true but do not properly infer the
conclusion.*

Science depends on rational discourse, based on the consideration of objective evidence, to arrive at a consensus of
what constitutes truth in any given situation. Unfortunately,
error can creep into any discussion, sometimes intentionally
but often by accident, if logical fallacies are committed. In this
discussion we will take a look at the different classes of logical
fallacies; our goal is to help you develop the facility of critical
thinking.

Persuasion, debate, and argument can take many courses,
but only some of them reliably lead to scientific truth. Someone may, for example, try to rationalize a position. But ratio-
nalization taken wrongly ("rational lies") implies that one is
either collecting data, or presenting evidence, to support a

predetermined position, which is nearly always bad. The position supported is determined by factors other than the evidence presented. There is also *propaganda*, i.e., statements made "... by those who want to persuade others to believe them." Propaganda may or may not be true, or it may be partly true, but to the propagandist that consideration is usually irrelevant—the only issue is whether or not it is believed. The claims of advertisers and marketeers sometimes fall into this category, but those of scientists *never* should. One need only look at the racial theories of Nazi Germany for an example of science perverted on a propagandist's altar.

The normal way that honest people try to persuade each other is by the process of rational dialogue. In this process, a person presents an argument consisting of all available relevant facts, propositions, and premises, including those facts that argue against their own position. It is through this process that scientific truth is wrested from nature's jealous grip.

The word *argument* in this context does not imply anger. It is used in the formal logical sense to mean reasons offered in support of, or opposition to, a position. To argue is to mutually consider both the pros and cons of the position under examination.

If the facts are true, the premises are valid, and both are pertinent to the issue at hand, then the argument necessarily implies the right conclusion in the minds of rational, unbiased hearers. At least, that's the ideal situation, which usually happens only when the argument is not burdened by emotional or political considerations. But arguments can get into serious trouble because a "fact" is really untrue, or the fact is taken out of context, or because one or more of its propositions are based on false premises, or because the propositions accepted don't logically lead to the desired conclusion.

Sources of Fallacy

Fallacy can arise from several sources. First, there is old-fashioned deceptiveness (fraud). The intent of the persuader is to convince you of the truth of some argument regardless of whether or not it is actually true. A necessary ingredient for a fallacy to be actual fraud is that the persuader knows that it is either a false argument, or that some of the "facts" are actually not facts at all (perhaps embellished—"made better"—or even embroidered out of thin air).

Another source of fallacy is old-fashioned fuzzy thinking. Perhaps the persuader has not fully thought out the theory, and is attempting to sell the idea with too little preparation. Prior to attempting to persuade another, you must first exercise a bit of forethought about the matter at hand. The key concept here is clear, focused, critical prior thinking about the problem, rather than free-floating, random musings.

The term "critical thinking" carries unfortunate emotional baggage for some people because of its negative connotations. In this sense, however, critical thinking refers to the open-minded attempt to analyze data, perceive patterns or trends within the data, evaluate the meaningfulness of the data, examine the meanings of the words used to describe the problem or define the data, and to then carefully reason from a sound basis in order to arrive at a conclusion or solve the problem.

One commentator said that "...the most important thing about critical thinking is how to defend your position." Unless an opinion can be articulated, orally and in writing, it can't be adequately defended. *A scientific opinion that cannot be defended might as well not exist.*

Still another source of fallacy is incorrect use of language. Emotion, or the necessity to win over or "one-up" the other person, or some other irrelevant factor may also lead to false reasoning. Regardless of the source or the honor of the persuader, it is important to guard against fallacies.

The first step in guarding against fallacies is to understand the types of fallacy that are most often encountered. Logician Irving M. Copi in his *Introduction to Logic, 7th Edition*, divides logical fallacies into two main categories: *fallacies of relevance* and *fallacies of ambiguity*. The former fallacies occur when the "... premises are logically irrelevant to, and therefore incapable of establishing the truth of, their conclusion." Fallacies of ambiguity are rooted in either ambiguous words that have multiple meanings, ill-defined meanings, meanings that are inappropriate for the issue at hand, or where erroneous grammatical construction renders an otherwise good argument erroneous.

Types of Fallacies

Logicians divide fallacies into several formal categories according to type. Although there is no truly "official" list of formal fallacies, and various experts divide them somewhat differently, there is general agreement at least on the main classes. (In addition to Copi, another source for readers interested in delving into this topic is Diane F. Halpern's *Thought and Knowledge: An Introduction to Critical Thinking, 2nd Edition*. Below you will find the types of fallacy that are often encountered. The categories overlap somewhat, so the classification scheme is not perfect.

Fallacies of Relevance

The matter of relevance is at the very heart of effective scientific argument. To be relevant, a fact, premise or proposition must have some pertinence to the issue at hand. Furthermore, it must have a direct and significant bearing on the subject matter—it provides evidence either for or against it. To be relevant, the argument must be based on logical connections, identifiable through reason, and not on the psychology of mere

emotion. In the sections below you will find a number of fallacies of relevance that are commonly encountered.

Type I: Argument from Intimidation

This fallacy is also sometimes called the *argument by appeal to
force* or the *put-down fallacy*. It is rooted in the view that "might
makes right," and dismisses any arguments to the contrary.
There are several situations where the Argument from Intimidation arises. One occurs when the arguer knows that his case
is weak, but for one reason or another persists in it. The Argument from Intimidation is often used when other, more rational,
means fail. If an argument based on provable facts, and demon-
strably correct premises, somehow fails to achieve its goal, then
the person may unethically resort to the Argument from Intimidation. It is also used when a person knows that their own
position is no more valid than the opposing position, so appeals
to a threat to achieve the goal that reason failed to secure.

Author Ayn Rand provided an example of the Argument
from Intimidation in several of her writings. (If interested,
look for *For the New Intellectual, The Virtue of Selfishness*, and
numerous issues of Rand's periodicals *The Objectivist* and *The
Objectivist Newsletter*.) This form of intimidation depends not
only on the actual words said, but also on body language and how
the words are uttered. The scenario goes something like this:
When a student challenged a professor whose position apparently did not bear scrutiny, the professor adopted a sarcastic
attitude, adjusted his *pince nez* glasses down onto the front of
his nose, leaned forward on the lectern in a threatening manner,
and using a very snide, patronizing voice said: "Surely, Ms.
Anderson, IF [contemptuous emphasis!] you had actually
READ [more contempt] the April 1912 edition of [voice rises
in level] *THE RADICAL INTELLECTUAL JOURNAL OF
IRREPLICABLE SCIENTIFIC MINUTIAE* [voice returns to
normal, cynical level], then you would have known that your

childish position had long ago been discredited by all reasoning people! [finished sentence with a disdainful, patronizing look on his face, and more than a hint of spittle in his voice]" Implied in this threat to the competence and dignity of the now sheepish student—whose semester grade and entire professional future as a scientist were at stake—was that she was not a reasoning person and was somehow derelict in her scholarly duty by not having read an ancient edition of an obscure academic journal of doubtful merit.

A good word for those who attempt to "convince" others by intimidation is *dinosaurbrains*.

Type II: Argument ad Hominem

The Argument *ad hominem* is always directed at the opponent, not the opponent's position. A lawyer related to me that a three-fold principle exists in the practice of law: *If the law is in your favor, then stand on the law; If the facts are in your favor, then stand on the facts; If neither the facts nor the law are in your favor, then attack the motives of the other side or the personal character of the other attorney.* The tactic in the Argument *ad hominem* is to attack the other person, not his or her position. That is, when one cannot disprove the assertions of the other side, then obscure the issue by making a personal attack on them. There is an unfortunate tendency among scientists to use *ad hominem* attacks when their own prior research is brought into question. One character stated that attacks by scholars on each other are so vicious because so little of real value is at stake!

Another form of *ad hominem* fallacy is sometimes called the *Genesis Fallacy* because it imputes error to an argument because of the origin of the argument, rather than its internal content. This fallacy is built on an emotional appeal, rather than on fact. For example, if a union organizer makes a proposal, then to management it must be a trap simply because it was "labor" who originated the proposal—its merits notwithstanding. And the other side of the coin is frequently seen too: A proposal must either be wrong, or founded on some sneaky, unspoken ulterior motive, simply because it was management who advanced it. In neither case are the merits of the proposal examined—only the originator.

Another type of *ad hominem* fallacy is sometimes called the *Circumstantial Fallacy*. That is, some special circumstance of the other person is sufficient reason to accept or reject a proposition. An example is the assertion that a certain man must accept the proposition that the national research budget must go higher next fiscal year simply because he is a research scientist.

A variation on this theme seen in science is to give special meaning to a person's affiliations. An argument may be doubted, not because it is false, but rather because of the natural interests of the person who makes it. The claim may be advanced that the person's self-interest so pollutes their reasoning that their argument must therefore be false. For example, try convincing an eco-activist of the reasonableness of a proposal from a motor car manufacturer to delay implementation of stricter clean air standards because of serious economic impact, or because science and engineering have not caught up with our wishes and desires for a cleaner environment.

These fallacies include a class in which one's associations are held to impute goodness or badness to either oneself or one's position. A medical scientist must be biased because he or she is paid by the drug company, and takes vacations on the company's yacht. This is the old "guilt [or virtue] by association fallacy." Issues of conflict of interest and propriety can properly be raised, but *criticism of scientific positions may be based only on the merit of the evidence presented.*

Defending against the *Argument ad hominem* requires the integrity to consider only the relevancy of the facts, the strength of the argument, and the context in which they naturally fit, rather than the emotional responses that come so easily.

Type III: Argument from Ignorance

This fallacy arises from the attempt to prove or disprove a proposition on grounds of the apparent nonexistence of opposing evidence. It is characterized by the attitude that "no news is good news." There are two main errors pertaining to the Argument from Ignorance, and they are based on the same flaw:

1. A proposition is true simply because one cannot prove that it is false, or

2. A proposition is false simply because one cannot prove that it is true.

The fatal flaw for either case is that we are asked to accept a negative... and negatives are, by their very nature, impossible to either prove or disprove. We may not have adequate evidence either for or against a proposition, which could mean that we are merely not ready to make a decision as of yet. Either the proper evidence has not yet been uncovered, or our present methods are not yet capable of finding the required evidence (often a problem in engineering development and scientific research at the frontiers of knowledge). It might also mean that we have simply not yet asked the right question of Nature. After all, an answer cannot be reached if the question has not been asked, or if the question was misframed at the outset.

There are three generally recognized exceptions to the Argument from Ignorance fallacy: *criminal law*, *qualified observer*, and the *standards of evidence* used in civil law. The first of these involves criminal court procedure in the law. The American justice system assumes that a person accused of a crime is innocent until proven guilty in a court of law; the burden of proof is on the accuser. The defendant may actually be quite guilty, and emotionally it seems imprudent to let him or her go free. But as a matter of public policy, to protect a greater value (freedom from a police state), we by convention require the state to prove guilt beyond a reasonable doubt. Thus, absence of proof of guilt is, by definition, a finding of "Not Guilty."

Second, there is a class of events that are deemed sufficiently special that any qualified observer would have noted them had they actually occurred. Given the existence of a qualified observer, and the fact that the observer did not note the occurrence of the event, then it is good evidence that the event did not, in fact, occur. A qualified cardiologist (heart doctor) examining an electrocardiograph (ECG) recording for a pathological heart rhythm anomaly may find none. That finding is good evidence that none existed. The opinion of a medical layman, or even other types of physician in some cases, has no merit in that situation. You and I could examine ECGs for a week and not form a valid opinion no matter how good our collective wisdom.

Third, according to the Rules of Evidence used in some civil courts, if an event is normally recorded as it occurs or is performed, then the absence of an entry in the pertinent record is good evidence that it did not occur. An example is found in the field of preventive maintenance (PM) of medical equipment. In a hospital, a very intensive PM program was in effect under which an inspector was required to record each inspection of certain electronic equipment in a log. No data was recorded other than the fact that the inspection took place, and the equipment was either found to be normal or sent to the repair shop. During a malpractice suit, the court ruled that the critical inspection had not taken place despite the testimony of the inspector to the contrary. Even though other evidence suggested that lapses in the recording function were common, the court said that the absence of a record is sufficient evidence that the inspection did not in fact occur. Such a position is less a fallacy of logic, however, than an operational policy under which the court believes that it must operate for practical reasons.

Type IV: Argument from Pity

This type of argument is used to evoke an emotional or sentimental response in order to get an argument accepted, despite the facts of the case or the logic of the premises. A number of examples of the Argument from Pity are available. First, there is the courtroom ploy. A defendant may be obviously guilty to all in the jury, and the defense attorney knows it. He may decide to appeal to the juror's sense of pity by describing the horrible circumstances of the person's life as the root of the crime. It sometimes works.

Another example was the turning point in a medical malpractice suit when the paraplegic plaintiff took the witness stand. She wheeled herself up to the raised deck where the witness chair was located, and then—wheezing audibly—she slowly and painfully pulled herself up to the chair as the whole courtroom watched. Faces in the jury showed distinct looks of concern and pity. It is significant that her testimony was little more than a *pro forma* description of her ill health. The reason that the attorney placed her on the stand was not that she had something to add to the testimony, but rather it was in order to let the jury see for themselves her pitiable state.

Type V: Argument by Appeal to Popular Opinion

The argument by public opinion is an emotional argument based on the *vox populi* (voice of the people) principle to win acceptance of a conclusion that is unsupported by facts, sound premises, or evidence. Appeals by advertisers, demagogues, propagandists, and some politicians fall into this category. It is also used to put pressure on regulatory authorities to accept quack medical treatments (e.g., the drug Laetrile) in the absence of scientific evidence, or in the face of contrary evidence. As long as one can whip up a large outcry of public sentiment for the proposal, then it is not necessary to actually prove the case.

The response of the public sometimes seems almost automatic. One suspects that a kind of "critical mass" effect exists in which an idea takes hold of the popular imagination once a certain number of the right people adopt it for their own. After that, a large number of people will suddenly adopt the idea—and it becomes the *de facto* "public view."

This form of argument often attempts to draw either positive or negative associations between the product or concept being sold and some greater value. In fact, such associations are probably key to success in some forms of persuasion. The classic examples are found in popular politics and in the advertising of consumer products. A politician may couch an argument for a bill in terms of patriotism, Mom, flag and apple pie (or against man-eating sharks, poison apples, and athelete's foot), rather than on the bill's actual contents.

A special form of the Appeal to Public Opinion (sometimes seen in connection with scientific facts) is the argument that "it's common knowledge" or "everyone knows it." In most cases such alleged common knowledge reflects not the universal acceptance of a self-obvious fact (as implied), but rather that the persuader dearly wants you to accept this "fact" or opinion but has not done sufficient homework to be able to convince you by other means. Even if large numbers of people in a society believe that something is true, it is no guarantee that it is, in fact, true.

In a subtle way, the *vox pop* argument can become a form of the Argument from Intimidation if there is an implicit "... so why don't you?" at the end of the sentence. If "everyone" or

> ### Selling Her Lips
>
> The advertiser is not selling fine clothing but success in business and romance. It is not the costly perfume that is offered, but rather an improved social life. And it is not the steak that is sold, but the "sizzle" that implies a pleasant dining experience. An author who wrote a book critiquing the advertising business took her title from a remark by her small child while watching a TV commercial: "Mommy, are they selling her lips?" In a way, yes, they are "selling her lips."

even just "most people" knows that something is true, then there must be some terrible deficiency in you if you don't also know it! Always be suspicious of the "common knowledge" ploy to reject it out of hand, and force your opponent to either prove the case or rescind the argument. Science has as little use for "common knowledge" as it has for "common sense."

Type VI: Argument by Appeal to Authority

An argument may either be so complex or so esoteric as to require the opinion of an expert witness. For example, you could hardly dispense medical advice unless you are a qualified physician. Similarly, you may have distinct opinions about why a bridge collapsed, but the opinion of a licensed professional civil engineer, who specializes in structures of bridges, is far more relevant. In these kinds of situations, the Appeal to Authority is valid.

There are, however, cases where the Appeal to Authority is not valid. Some people pull "scientific studies" or "expert opinions" out of thin air without so much as a shred of evidence that they either really exist or are properly interpreted (not an easy task in many cases!). In my job I often hear of a contractor's engineering or scientific analysis that supports their position. But when I ask for a copy, the analysis is often either shallow and self-serving, or does not really conclude what was represented previously, or in reality doesn't even exist! People who complain about writing footnotes or endnotes in a paper should realize that they are the validation of an Appeal to Authority, and without those notes such appeals are always suspect.

Computers represent a new form of Appeal to Authority, but in this case it is the all-powerful computer that is accepted as the "authority." Admiral Grace Hopper returned to active duty with the U.S. Navy after retiring from a successful career as university professor, computer expert, and co-inventor of a major computer language (COBOL). (Admiral Hopper gives the audience a "take away" from her lectures. Her aide gives out pieces of wire that are about 11.8 inches long—the distance light travels in a nanosecond. Her "nanoseconds" are valued souvenirs of her lectures.) When Hopper first came back to the Navy, she was tasked to prepare a budget for her newly formed organization. After due thought, Hopper scrawled the figures on a sheet of yellow legal paper and presented them to the budget officer. Aghast at the informality, the budgeteer summarily rejected them out of hand. Undaunted, Hopper returned to her office and typed the same numbers, in the same format, into a computer. She printed them out on standard "green bar" computer paper, and then returned to the budget officer. The figures "generated" by the computer were accepted immediately. They carried more weight than exactly the same information hand-written! Same data, different format. In this case, appeal to an electronic digital "authority" outweighed the same information presented by a genuine world class human authority!

Computer modeling and simulations are common today, especially in science and engineering. These simulations represent a lower-cost way of solving real problems, especially in the early phases of a research project. The results of mere simulations often take on an aura of great authority, even though they may be really little more than educated guesses. In one field, radar engineering, the same computer model for a generic radar predicted detection ranges of 29 miles, 51 miles, and 77 miles depending on the assumptions made when the operating parameters were input. The simulation is a tool, not Absolute Truth, and must be continuously refined as assumptions become hard knowledge. Even validated models are highly contextual, and their results must be treated as tentative.

In other cases, we find that the authority appealed to is not a truly valid authority for the specific issue. Ranking officials, scientists (especially Nobel Laureates), professors, and persons who are experts in professional fields are often looked up to for advice on matters outside of their valid area of expertise. One might cite the opinion of a medical doctor on events in the Middle East, for example, as if the M.D. degree somehow conferred universal expertise. (Some MDs think that it does!) In that instance, the credential is not valid for the matter under discussion, and the good doctor's opinion is no more valid than that of any other educated person.

Type VII: Argument from Accidental Circumstance

This fallacy involves erroneously applying a general rule to a particular case that contains accidental circumstances that render the general rule invalid. It does not recognize that many times "situations alter cases." It is particularly dangerous when narrow, pedantic legalists or entrenched bureaucrats attempt to mechanically appeal to a fixed set of rules to decide complex issues, especially when "scientific facts" or "laws" are cited outside of their area of validity. The base error often involves acceptance of a rule as universal when, in fact, it is not applicable to many specific situations.

One sometimes sees a situation in older scientists who are wedded to the past, especially to theories that they helped initiate. I once asked a scientist about the application of chaos and fractal theory to his area, and he snorted "It's a fad, nothing to it." He never did accept the verdict of his colleagues when successful fractal and chaos papers started showing up in the journals, and changed the field considerably... maybe he was a dinosaurbrain.

Type VIII: Argument from Converse Accident

This form of fallacy derives from erroneously accepting the results of sampling investigations as applicable to all cases. We may seek to characterize a large class of cases from a too-small sample. The sampling method produces error if the sample is either atypical or if it is too small. Good sampling technique requires three things: (1) a sample that is large enough to be statistically significant, (2) that the samples be typical of the whole population, and (3) that they be selected randomly (every member of the set has an equal chance of being selected) in order to suppress personal and other biases.

One often sees this error made with respect to lay interpretations of scientific research data. One key area where examples abound is popular medical books and articles. Often based on research that is highly tentative (and understood as such by competent physicians and medical scientists), articles by lay writers sometimes optimistically generalize about research that may later be either totally invalidated by further research, or found to apply only to specially selected populations.

The popular habit of taking an aspirin a day in order to prevent heart attacks is an example of too-quick generalization from scientific literature. The original research, part of a nationwide program called the Aspirin Myocardial Infarction Study, showed an encouraging high correlation in *males* of decreased incidence of *second* heart attacks and taking an aspirin *every other day*. Some over-the-counter drug distributors took to packaging *daily* aspirin doses on calendar blister cards, and advertised for *all people* every day. What was missed in much of the popular literature was that the patient population was highly selected (for valid scientific reasons), and that the dosage and frequency were half what was advertised by the aspirin makers. Also overlooked was that some patients with a sensitivity to aspirin, bleeding problems (e.g., ulcers) and other medical problems should not take this path. The false popular generalizations about the study inclined people to self-medicate in an area where they should have consulted their physician for a qualified medical opinion.

Type IX: Argument from Selected Instance

The Argument from Selected Instance involves narrowly selecting evidence to support a theory, while ignoring contrary evidence. It is very easy to fall into this trap because people are usually excited about their own theory. If they are not careful, they will collect facts that support that theory and overlook or reject facts that are inconvenient. But scientific method, and indeed all rational thinking, requires one to consider all relevant facts before forming a position or theory. It is also part of scientific ethics to present (and answer) all contrary evidence that either supports other, competing, theories, or contradicts the theory under investigation.

A way to guard against the Argument From Selected Instance is to use multiple sources of information from both sides of the issue. It must be noted that no source is truly objective, but with a sufficient number of biased inputs on both sides of an issue, one can often cancel out the biases and arrive at some reasonable approximation of the truth.

Type X: Argument from False Cause

This fallacy is also sometimes called the *post hoc* error. It is often falsely assumed that a relationship in time between events implies cause and effect. If event B always follows event A, then A must be the cause of B. Such a generalization is valid only if there is other evidence of causality than the simple time relationship. But there are many other reasons than causality why events might be correlated in the "B follows A" sense. It may be purely a coincidence. Or, it may be that both are caused by a co-factor that affects "A" a little quicker than "B." If I get the flu today, and you get it next week, and there was no contact between us anytime around the time I was sick, then it cannot be reasonably concluded that my flu caused your flu. However, if there is evidence that we were housed together for several hours in a small room with poor ventilation, shared eating utensils, and shook hands (a major transmission path for such diseases!), then a reasonable and prudent person may assume that my flu had something to do with yours — especially if the incubation time for the disease is close to the time period between our contact and the onset of your illness.

Traffic signals show a progression in which red lights always follow yellow lights. Can we therefore conclude that red lights are caused by yellow lights? Or is there another factor, like a stepper relay or solid-state controller that causes the lights to turn on and off in sequence?

Suppose you were to plot a graph of both the incidence of armed robberies and the number of marriages in your county for the years 1900 to the present. It is clear from such a graph that the number of armed robberies and the number of marriages increases at the same time. Can we conclude, therefore, that marriages cause armed robberies? Or is it that armed robberies cause marriages? The simple truth is that both are related to population growth, which is an uncharted third factor. To assert either proposition above is to fall into error.

Type XI: Begging the Question

There are several forms of this fallacy. First, "Begging the Question" is looking for a premise that will support making the proposition at hand into a supposedly valid (if preconceived) conclusion. That is, narrowly searching for data to support a position, while rejecting data that does not support it. Second, accepting as a premise the very conclusion that it is supposed to prove. Third, using the same propositions as both premises and conclusions. This is the classical "circular argument." That is, when the reason is linked logically with the proposition, the result is a circle.

For example, consider the statement "we need to raise prices because the prices are too low." In this instance, we are asked to accept higher prices for the reason that the prices are too low. Compared to what? A reason for raising prices might be that costs are up, or production quantities down, so that the manufacturer can no longer make the product within the guidelines of the agreed profit margin. But to assert that the "reason" for raising prices is that they are too low is to beg the question.

Type XII: Oversimplification of a Complex Question

The oversimplification fallacy consists of demanding or using a binary "yes or no" answer for a complicated question for which there is no simple answer. This fallacy is easy to fall into in experiments where hypothesis are framed to require binary answers. Consider the question: "Have you stopped beating your wife?" This question is inherently erroneous because it implicitly assumes that a prior unasked question ("Do you beat your wife?") is already answered in the affirmative. "Yes" or "no" implicitly answers the unasked question while explicitly answering the asked question. For that reason, it is inherently tricky and always logically dangerous to answer complex questions with a simple binary response like "yes" or "no." This fallacy can arise in science when we set up experiments to answer a complex question "yes/no" or "true/false." Either a series of progressively more refined experiments should be done, or the question asked of nature needs to be better isolated and simplified (if that is possible).

Type XIII: False Dichotomy Error

The Argument by False Dichotomy is another binary fallacy, and is therefore similar to the complex question fallacy above, but does not require a "yes" or "no" answer. A health systems computer salesman once told me that he doesn't really give his customers the choice that they thought they were getting in the selection of nurse's workstation hardware. He might ask: "Do you want a VGA or VGA-Plus terminal?" That is a binary decision that overlooks the fact that there are a number of cheaper color and monochrome computer display terminals on the market. He completely ignored the fact that VGA monochrome displays—at one-fifth the price of color VGA terminals—were just as useful, and are in many ways superior in those situations where graphics do not have to be in color. He also failed to tell the customers that the ordinary monochrome displays (with limited graphics capability) cost one-tenth as much as VGA color. But then again, it wasn't serving his customer's best interests that was on his mind.

The false dichotomy fallacy is also sometimes called the *suppressed information fallacy* if it rests on keeping critical information from critics. By denying others access to critical information, one can achieve an advantage that would otherwise not be available were the truth known.

Guarding against the false dichotomy requires you to examine unspoken assumptions or qualifiers, as well as hidden premises or incomplete factual statements. You must also consider the unmentioned alternatives or novel combinations of the elements of the argument that are not specifically mentioned by the other side. Much of the information that does not surface in the discussions is what is really at the base of a false dichotomy, and it is those suppressed elements that you must seek.

Type XIV: Ignoratio Elenchi Fallacies

This fallacy involves attempting to prove or disprove a proposition by either proving or disproving some other proposition that is not at issue—that is, attempting to argue for a conclusion by appealing to another conclusion altogether. Some fallacies of this class are emotionally based, but that factor is not necessary for the definition to apply.

Type XV: Argument from the Stolen Concept

Another common form of fallacy is the Argument from the Stolen Concept. The Stolen Concept involves the transfer of a concept into a context or frame of reference where it does not logically belong. Some people see this fallacy as a special case of Ignoratio Elenchi. The Stolen Concept fallacy is seen whenever someone attempts to impute either virtue or vice from one concept to an unrelated concept. The Stolen Concept is not committed simply because analogy or metaphor is used (which, in context, are valid means for communicating some truths), but rather when these are taken too far.

Stolen concepts are frequently heard among callers to radio talk shows who attempt an emotionally based transference of the issue at hand, or at least its name, to their own pet issue. For example, both sides of the abortion issue are guilty of this fallacy. The "anti's" will liken abortion to the Jewish Holocaust in World War II Europe, while the "pro's" will liken restrictive abortion laws to chattel slavery of women or government oppression of the downtrodden. In the minds of rational, critically thinking people, neither side does their position any real good by using stolen concepts. But since most people are not critical thinkers, it is unlikely that this fallacy will disappear from public discourse.

Type XVI: Slippery Slope Fallacy

This type of fallacy asserts that allowing a small concession will necessarily result in an inevitable downward slide towards additional—far more serious—concessions. A metaphor for this type of fallacy is the "camel's nose proverb." Supposedly, if a camel is permitted to put his nose under the tent, it is only a matter of time before the entire camel is inside the tent. "Give 'em an inch and they'll take a mile" typifies this fallacy.

Type XVII: "Straw Man" Fallacy

The straw man fallacy involves restating or reframing an opponent's argument in order to knock it down. The idea is to implicitly assert that the distorted argument is actually the original argument, so that a rebuttal to the weaker "straw man" effectively counters the real argument.

A straw man argument might be used to counter an assertion or proposal that you make. For example, consider the statement that "...working scientists should be consulted by the Administration when making decisions affecting the organization." Although consistent with modern management concepts, and despite the fact that it might broaden management's resources in the decision-making process, an opponent of the idea might counter by rephrasing your argument into "workers should manage the organization." They then shoot down that proposal, rather than the original, less-radical proposal. They may assert that the original proposal is thereby nullified, with a statement such as "...it is managers, not workers, who are charged with the responsibility for managing the organization." That statement cannot be challenged successfully because it is true. But the statement is also irrelevant to the original statement, and therefore should not be allowed to stand as an argument against it.

Another form of the straw man argument is the *Argument by False Analogy or Metaphor*. When the other side attempts to redefine your position in terms of parables, analogies, metaphors, or other literary devices, be wary! These devices are quite useful in making complex issues clear, especially those ineffable sorts of issues for which language so often fails. There is, however, a distinct possibility that an initially appealing literary device may lead to a straw man situation, rather than serving to clarify complex issues. One must guard against reframed arguments, incomplete statements of a position, or an out-of-context argument, in order to prevent loss of a valid position through this form of fallacy. It is worthwhile to be very wary whenever an opponent reframes, restates, or in any way "clarifies" your position. The words selected for the restatement may well weaken your position without your being aware of it.

Type XVIII: Incomplete Comparisons

A false argument is sometimes made by not informing the other party of all of the pertinent facts. How many pain relievers do we hear are better than competing brands? Nearly all of them make the claim! One may hear the claim that Brand-Z is superior because ". . . no pain reliever was found better than Brand-Z." When a magazine reporter ran down the source of the claim, it was found that the "pain other than headache" was intractable cancer pain, and that all of the pain relievers tested produced approximately the same relief. . . including a "sugar pill" placebo. True, no pain reliever was better than Brand-Z, but then again, Brand-Z wasn't better than any of the others—the unreported facts caused an incomplete comparison.

Type XIX: Argument from Tradition

The Appeal to Tradition is exemplified by the attitude, "That's the way we've always done it." Tradition has an alluring appeal, and should not be dismissed out of hand, but is also the refuge of dinosaurbrains. It tells us what worked once before, and that can be a valuable lesson, but tradition can also be both a trap for the unwary and a crutch for the lazy.

The problem with the Appeal to Tradition is seen when tradition is invoked mindlessly. It may well be that an old way of doing business got that way because it is sound. But, unexamined, an old way of doing business can ossify a person who is faced with continuously changing reality. Change is a constant of life, and is present even when nothing else remains constant. The only proper way to deal with changing reality is to continuously examine, continuously verify, and, where needed, continuously modify the system.

There is one use for the Appeal to Tradition that is quite valid. By making such an appeal, the open-minded person can force advocates of change to prove their case. Change run amuck is chaos, not improvement. The idea is to set up a creative tension between the advocates of change and the advocates of tradition, and then arrive at truth as they correct each other.

Type XX: Argument by Anecdote or Testimonial

One form of argument uses testimonials or anecdotal data in order to convince. For example, some over-the-counter pain reliever advertisements feature an actor portraying a fellow sufferer who proclaims that the Zorch brand of aspirin relieves his or her headaches and arthritis better than other brands. Testimonials and anecdotal data are also used extensively in support of health foods, vitamin and mineral supplements, and other health-related items. But these forms of "data" are not scientifically valid, since they are subjective. The only reasonable basis for such claims is a properly validated scientific study performed by disinterested parties...and such studies are often lacking in advertising claims. ("Disinterested" means "has no stake in." It is not the same as "uninterested.") It is their uncontrolled nature, and the fact that results are often reported by unqualified observers using highly subjective, ambiguous, or variable criteria, that makes testimonials and anecdotal data difficult to accept.

Anecdotal data may provide what cops and Supreme Court justices call "probable cause" for committing some resources in a preliminary exploration of a scientific study, but do not offer sufficient proof for a conclusive statement.

Fallacies of Ambiguity

The fallacies of ambiguity are generally fallacies of clearness. The fallacies of ambiguity generally arise from using (1) words that have multiple meanings, (2) words whose meanings shift subtly in the course of an argument, or (3) sloppy grammatical construction.

Type XXI: Fallacy of Equivocation

The fallacy of equivocation sometimes results from using words that have multiple meanings. In an engineering specification the word "nominal" was used in regard to a length (e.g, "nominally 50 feet"), without regard for either a well-understood definition, standard convention, or some sort of agreed-upon operational definition. The word "nominal" could legitimately mean any of the following:

1. A word or word group functioning as a noun.

2. Constituting or bearing a name.

3. Approximate, designated, or theoretical size from which real members of the class might deviate.

4. Existing in name only, not in reality.

The contractor, obviously, would prefer numbers (3) or (4) from the above list! If number (3) from the list above is accepted, then "50 feet" could mean "50 feet ± (tolerance)." Is the unstated tolerance 1 inch? Is it 1 foot? Is it 12.6 inches? Virtually any length anywhere close to 50 feet will satisfy the requirement because of the inherent ambiguity of the word "nominal."

Or consider the simple word "may." It could mean the *fifth month of the year on the Gregorian calendar*, or it could indicate *permission from a superior to take some action*, or that a certain *expectation based on a probability that the event will occur*, or *have the ability to* (same meaning as "can"), or *have liberty to*. One thing that is plain about the word "may" is that its uses are not always plain. Such simple, one-syllable words are often pregnant with unexpressed or implied meanings. Lawyers love 'em.

Type XXII: Fallacy of Ambiguity

This error arises from a premise that is formulated based on incorrect grammar, ambiguous grammar, or clumsy construction. There was a large high school auditorium in our area which had a lobby that rivaled some theaters in the area. In those days, before the Surgeon General's report, adults and authorized students were allowed to smoke in certain designated areas around the auditorium during intermissions of plays or musical performances. A sign at one end of the lobby attempted to create a smoke-free zone for non-smokers (25 years in advance of that ethic becoming widespread). The sign read: NO SMOKING IS PERMITTED UNDER THIS SIGN. Does the sign mean that one is permitted to not smoke under that sign, but must smoke everywhere else in the building? Also, what does "under" mean? Between the back of the sign and the wall? Does "under" mean some undefined three-dimensional wedge of XYZ space somewhere in the vicinity of, and below, the sign? The sign was not that clear in its meaning, even though most reasonable people would probably understand its intent.

Type XXIII: Error of Accent or Emphasis

This fallacy is a deceptive argument that is based on a change or subtle shift of meaning of words. The meaning of the argument may depend on how it is presented, emphasized, or accented. This fallacy depends on the implicit acceptance by the hearer of the argument, without critically reviewing the actual meanings of the words used.

In spoken arguments the accent or emphasis can be placed on the wrong word, or body language be adjusted to distort the meanings of the words. In written arguments, emphasis can be added by underlining, italicizing, bold facing, or otherwise highlighting the words that are intended to stand out. Another way that words can be distorted is to lift them out of context or report them incompletely.

Still another error of accent or emphasis is fine print, or the relegation of critical information to endnotes. A new car dealer might advertise *Zorchmobiles* at $7,995 *and up*. But when you find the one $7,995 car in his stock (which legitimized the ad), you find that it is an undesirable, stripped-down model with no passenger seats, and it is effectively "nailed to the show room floor." All the other ". . . and up" cars in stock sell for more than $12,000.

Type XXIV: Error of Composition

This fallacy is derived from imputing the attributes of a single element to a collection of those elements which make up a whole. Even if all of the elements are the same, the attributes are not necessarily transferable to the whole. It is frequently the case that the whole may be more, or at least different, than the sum of the parts. Biology and medicine provide an example. One can observe individual cells of an organ and create a list of attributes that define that type of cell. Consider, for example, the cells of the heart wall. These muscle cells are long and stringy, and when stimulated they contract longitudinally. But when they work together in the heart, it is found that the organ is not long and stringy and when stimulated it beats with a twisting motion (once described in *Scientific American* as being like "wringing out a sponge")—not a longitudinal motion.

The basis for this kind of error is confusion over the distributive and collective uses of a term or word. For example, consider the statement that "2000-lb bombs are more destructive than 500-lb bombs." We know that one 2000-lb bomb usually makes a wider, deeper smoking hole in the ground than one 500-lb bomb. The statement is true when only one 500-lb bomb is compared with one 2,000-lb bomb. But if 2000 lbs of explosive are dropped in the form of four 500-lb bombs instead of one 2,000-lb bomb, then the total damage is likely to be greater even though the total weight of the explosives are the same. The individual smoking holes in the ground may not be as deep, but there are more of them spread over a wider area.

Type XXV: Error of Wrongful Division

This class of fallacy is the inverse from the previous type—it imputes to the members of a class the attributes of the whole class. Consider again our human heart example. The attributes of the heart myocardium are not the same as the attributes of the individual cells of the myocardium. It is not always the case that "what is true of the whole is true of the parts."

Type XXVI: Prose or Vague Statistical Comparisons

Persuaders are sometimes prone to using adjectives instead of real data in making comparison claims. One might hear, for example, that Brand-Z aspirin gives *faster* pain relief than other brands. Faster? Faster than what? Were all brands tested? How was "fast" measured, and against what standard? Or, what about the mouthwash that claims that it "cleans away more plaque than other brands." Which other brands? How much is "more"? How was "more" determined?

Some advertisers assert numbers in a claim because people hearing the ad will place more trust in numbers... it is assumed that these numbers resulted from a careful scientific study. Right? Does a mouthwash that cleans away "...72% more plaque" really meet that claim? And, 72% with respect to which competitor? Were all competing brands tested in exactly the same manner? Who recognizes the mouthwash-testing protocol other than its manufacturer's in-house or contract laboratory? Do independent experts also recognize the test method? Those are the kind of questions that often force some light onto vague comparisons.

Conclusion

Reason is a powerful scientific tool when done correctly. But when forced off track by fallacies, reason is perverted and truth suffers. More specifically, people and organizations can be harmed when critical decisions are based on fallacies; errors hurt. But when a person is skilled in recognizing fallacies, the probability of making significant errors is reduced considerably.

References

Irvine M. Copi, *Introduction to Logic, 7th Edition,* MacMillan (New York, 1986).

Diane F. Halpern, *Thought and Knowledge: An Introduction to Critical Thinking—2nd Edition.*

APPENDIX B

BASIC Program For Simple Statistical Analysis

SimpleStats is a BASIC computer program that acts as a data logger by permitting you to enter either single point (X_i) or ordered pair (X_i, Y_i) data, and then store it on disk using a filename of your choice. If you wish to add to a file, *SimpleStats* permits you to load the file, add data to it, and then restore it under the same or a different name. The program will make the standard statistical calculations on the entered data (minimum value, maximum value, range, variance, standard deviation). The results are normally stored on the screen, but an option allows you to print out a hardcopy of the data set and the results of the statistical calculations.

Executable diskette for MS-DOS machines available from the author. P.O. Box 1099, Falls Church, VA 22041-0099.

```
10    REM SIMPLE STATS (STATS 7) [4 August 1991]
20    CLEAR:KEY OFF
30    DIM X(1000):DIM Y(1000):DIM X$(100):DIM
         Y$(100)
40    DIM XX$(100):DIM YY$(100)
50    REM MAIN PROGRAM EXECUTION REGION
```

```
60    CLS
70    GOSUB 370:REM Go get opening subroutine
80    GOSUB 1340:REM Go get main menu
90    ON MENU GOSUB
      2900,5380,5930,180,7070,1720,10,130
100   IF MENU=8 THEN 130
110   IF MENU=2 THEN 190:REM Test to see if add to
      existing file
120   GOTO 80
130   REM ENDING SUBROUTINE
140   CLS:LINE (470,200)–(168,150),3,BF
150   LOCATE 13,28:PRINT "Goodbye, ";C$
160   TIMELOOP=TIMER:WHILE
      TIMER<TIMELOOP+1:WEND
170   CLS:SCREEN 0
180   END
190   REM Subroutine to test for adding data to existing disk
      file
200   CLS:LINE (570,260)–(100,120),3,BF
210   LOCATE 12,20:PRINT " Do you want to add to the
      retrieved file, or "
220   LOCATE 13,20:PRINT " return to the main menu?
      "
230   LOCATE 15,20:PRINT " (A)dd data to disk file
      "
240   LOCATE 16,20:PRINT " (M)ain menu
      "
250   LOCATE 18,20:PRINT " MAKE SELECTION
      ";:LN$=INPUT$(1)
260   IF LN$="A" THEN 310:REM Go to the data collection
      routines
270   IF LN$="a" THEN 310:REM "   "   "    "       "
280   IF LN$="M" THEN 360:REM Return to main menu
290   IF LN$="m" THEN 360:REM   "    "  "  "
300   GOTO 200
```

```
310   CLS:REM Go to either X or XY input routine
320   ADD=1
330   IF DC=1 THEN GOSUB 3200:REM Go to X-data
         collection routine
340   IF DC=2 THEN GOSUB 3900:REM Go to XY-data
         collection routine
350   GOTO 360
360   DCF=1:GOTO 80
370   REM OPENING SCREEN SUBROUTINE
380   NTE(1)=523.25:NTE(2)=493.88:NTE(3)=
         523.25:NTE(4)=587.33:NTE(5)=659.26
390   NTE(6)=698.46:NTE(7)=783.99:NTE(8)=
         880:NTE(9)=987.77:NTE(10)=1046.5
400   CLS:SCREEN 9:XXX1=400:XXX2=100:YYY1=
         50:YYY2=200:M=10:COLOR 15
410   LINE (XXX1,YYY1)-(XXX2,YYY2),,B:SOUND
         NTE(M),10
420   M=M-1:IF M = 0 THEN 440 ELSE430
430   XXX1=XXX1+10:XXX2=XXX2+10:YYY1=
         YYY1+10:YYY2=YYY2+10:GOTO 410
440   COLOR 14:LOCATE 12,32:PRINT "Simple
         Stats":COLOR 15
450   LOCATE 14,26:PRINT "Copyright 1991 J.J. Carr"
460   TIMELOOP=TIMER:WHILE TIMER < TIMELOOP +
         2:WEND
470   CLS:KEY OFF:SCREEN 9
480   DIM A(50)
490   FOR I = 1 TO 31:READ A(I):NEXT I
500   DATA 0.,0.65,1,2,3,4.5,6.5,9,12.5,16,20,23,27,29,31,32
510   DATA 31,29,27,23,20,16,12.5,9,6.5,4.5,3,2,1,0.65,0
520   COLOR 3:LINE (451,257)-(125,257):LINE (125,257)-
         (125,75)
530   COLOR 14:FOR I = 1 TO 31:SOUND
         200+(10*A(I)),.9
540   PSET (125+(I*10),250-5*A(I))
```

```
550   TIMELOOP=TIMER:WHILE TIMER < TIMELOOP +
         .1:WEND
560   NEXT I
570   B = 257
580   FOR I = 1 TO 9
590   COLOR 3:LINE (128,B)–(120,B):B=B–20:NEXT I
600   A = 125
610   COLOR 3:FOR I = 1 TO 17
620   LINE (A,259)–(A,253):A=A+20:NEXT I:COLOR 14
630   COLOR 2:LINE (286,259)–(286,90):COLOR 14
640   LOCATE 19,58:PRINT "X":LOCATE 20,37:PRINT "X"
650   LOCATE 19,15:PRINT "0":LOCATE  6,16:PRINT "P"
660   COLOR 14:LINE (286,262)–(294,262)
670   LOCATE 24,1
680   H = 145:V=290
690   LINE (H,V)–(H–20,V):REM Horiz. Line
700   LINE (H–20,V+15)–(H–20,V):REM Vert. Line
710   LINE (H,V+15)–(H–20,V+15 ):REM Horiz. Line
720   LINE (H,V+31)–(H,V+16):REM Vert. Line
730   LINE (H,V+31)–(H–20,V+31 ):REM Horiz. Line
740   H = 175
750   LINE (H,V)–(H–20,V):REM Horiz.
760   LINE (H–10,V+31)–(H–10,V+2  ):REM Vert.
770   LINE (H,V+31)–(H–20,V+31):REM Horiz.
780   H=190
790   LINE (H–5,V+31)–(H–5,V):REM Vert.
800   LINE (H–2,V)–(H+9,V+31):REM Slant right
810   LINE (H+22,V)–(H+11,V+31):REM Slant left
820   LINE (H+24,V)–(H+24,V+31):REM Vert.
830   H = 230
840   LINE (H–5,V+31)–(H–5,V):REM Vert.
850   LINE (H+17, V)–(H–5,V):REM Horiz.
860   LINE (H+17,V+14)–(H+17,V):REM Vert.
870   LINE (H+17,V+14 )–(H–5,V+14 ):REM Horiz.
880   H = 258
```

```
890  LINE (H,V+31)–(H,V):REM Vert.
900  LINE (H+15,V+31)–(H,V+31)
910  H = 303
920  LINE (H,V)–(H–20,V):LINE (H,V+15)–(H–20,V+15):
       LINE (H,V+31)–(H–20,V+31)
930  LINE (H–20,V+31)–(H–20,V):REM Vert.
940  H =365
950  LINE (H,V)–(H–20,V):REM Horiz. Line
960  LINE (H–20,V+15)–(H–20,V):REM Vert. Line
970  LINE (H,V+15)–(H–20,V+15 ):REM Horiz. Line
980  LINE (H,V+31)–(H,V+16):REM Vert. Line
990  LINE (H,V+31)–(H–20,V+31 ):REM Horiz. Line
1000 H = 392
1010 LINE (H–10,V+31)–(H–10,V+2  ):REM Vert.
1020 LINE (H,V)–(H–20,V):REM Horiz.
1030 H = 403
1040 LINE (H,V)–(H+17,V):REM Horiz.
1050 LINE (H,V+15)–(H+17,V+15):REM Horiz.
1060 LINE (H,V+31)–(H,V):REM Vert.
1070 LINE (H+17,V+31)–(H+17,V):REM Vert.
1080 H = 448
1090 LINE (H–10,V+31)–(H–10,V+2  ):REM Vert.
1100 LINE (H,V)–(H–20,V):REM Horiz.
1110 H = 480
1120 LINE (H,V)–(H–20,V):REM Horiz. Line
1130 LINE (H–20,V+15)–(H–20,V):REM Vert. Line
1140 LINE (H,V+15)–(H–20,V+15 ):REM Horiz. Line
1150 LINE (H,V+31)–(H,V+16):REM Vert. Line
1160 LINE (H,V+31)–(H–20,V+31 ):REM Horiz. Line
1170 TIMELOOP=TIMER:WHILE TIMER < TIMELOOP
       +2.5:WEND:CLS
1180 COLOR 7,1:COLOR 14:LINE (500,220)–
       (125,120),7,BF
1190 LOCATE 12,22:PRINT " Data Collection, Data
       Logging and "
```

```
1200 LOCATE 13,22:PRINT " Statistical Calculation
      Program for "
1210 LOCATE 14,22:PRINT " Amateur and Student
      Scientists "
1220 TIMELOOP=TIMER:WHILE
      TIMER<TIMELOOP+5:WEND
1230 GOSUB 2760
1240 REM Announcement subroutine
1250 CLS:SCREEN 9:COLOR 7,1:COLOR 14
1260 LINE (525,270)–(105,90),3,BF
1270 LOCATE 10,32:PRINT " SIMPLE STATS ":COLOR 15
1280 LOCATE 13,17:PRINT " This program collects data,
      saves data to disk "
1290 LOCATE 14,17:PRINT " and performs statistical
      calculations on the "
1300 LOCATE 15,17:PRINT " data. It will also perform other
      functions. "
1310 COLOR 14:LOCATE 18,25:PRINT " Press any key for
      main menu "
1320 A$=INKEY$:IF A$="" THEN 1320
1330 CLS:RETURN
1340 CLS:REM MAIN MENU SUBROUTINE
1350 SCREEN 9:COLOR 7,1:COLOR 14:LINE (500,270)–
      (150,60),3,BF
1360 LOCATE 7,25:PRINT "                         "
1370 LOCATE 8,25:PRINT " (I)nput data from keyboard
      "
1380 LOCATE 9,25:PRINT " (L)oad data file from disk      "
1390 LOCATE 10,25:PRINT " (S)tatistical calculations      "
1400 LOCATE 11,25:PRINT " [future expansion]            "
1410 LOCATE 12,25:PRINT " (H)ardcopy printout
      "
1420 LOCATE 13,25:PRINT " (G)enerate random number
      "
1430 LOCATE 14,25:PRINT " (R)estart program            "
```

```
1440  LOCATE 15,25:PRINT " (E)nd program                    "
1450  LOCATE 16,25:PRINT "                                  "
1460  LOCATE 18,32:PRINT " MAKE SELECTION
         ";:MENU$=INPUT$(1)
1470  MNU=VAL(MENU$)
1480  IF MNU > 0 THEN GOTO 1340
1490  IF MENU$="0" THEN GOTO 1340
1500  IF MENU$="" THEN GOTO 1340
1510  IF MENU$="I" THEN MENU = 1
1520  IF MENU$="i" THEN MENU = 1
1530  IF MENU$="L" THEN MENU = 2
1540  IF MENU$="l" THEN MENU = 2
1550  IF MENU$="S" THEN MENU = 3
1560  IF MENU$="s" THEN MENU = 3
1570  IF MENU$="D" THEN MENU = 4
1580  IF MENU$="d" THEN MENU = 4
1590  IF MENU$="H" THEN MENU = 5
1600  IF MENU$="h" THEN MENU = 5
1610  IF MENU$="G" THEN MENU = 6
1620  IF MENU$="g" THEN MENU = 6
1630  IF MENU$="R" THEN MENU = 7
1640  IF MENU$="r" THEN MENU = 7
1650  IF MENU$="E" THEN MENU = 8
1660  IF MENU$="e" THEN MENU = 8
1670  IF MENU < 1 THEN GOTO 1340 ELSE 1680
1680  IF MENU > 8 THEN GOTO 1340 ELSE 1690
1690  IF MENU = 0 THEN GOTO 1340 ELSE 1700
1700  IF MENU$ = "" THEN GOTO 1340 ELSE 1710
1710  RETURN
1720  REM RANDOM NUMBER GENERATOR
         SUBROUTINE
1730  SCREEN 9:COLOR 7,1:COLOR 14:DIM
         RANNUM(1000)
1740  CLS:RNG=0:RNG$=""
1750  LINE (540,260)-(135,120),3,BF
```

```
1760  LOCATE 11,20:PRINT "                              "
1770  LOCATE 12,20:PRINT " 1. Generate a single random
         number        "
1780  LOCATE 13,20:PRINT " 2. Generate a table of N
         random numbers      "
1790  LOCATE 14,20:PRINT " 3. Generate random numbers
         until told to quit "
1800  LOCATE 15,20:PRINT " 4. Return to main menu    "
1810  LOCATE 16,20:PRINT "                              "
1820  LOCATE 18,35:PRINT " MAKE SELECTION ";:
         RNG$=INPUT$(1)
1830  IF RNG$ = "" THEN 1740
1840  RNG=VAL(RNG$)
1850  IF RNG < 1 THEN 1740
1860  IF RNG > 4 THEN 1740
1870  IF RNG = 4 THEN 2740 ELSE1880
1880  ON RNG GOTO 1890,2050,2430,2740
1890  REM Subroutine for single random number
1900  CLS
1910  LOCATE 12,25:PRINT "Highest number in range:";
         :INPUT RANDOM$
1920  IF RANDOM$="" THEN RANDOM$="1000"
1930  RANDOM=VAL(RANDOM$)
1940  IF RANDOM=0 THEN 1720 ELSE1950
1950  IF RANDOM=1 THEN 1960 ELSE1990
1960  A%=VAL(MID$(TIME$,7,2))
1970  RANDOMIZE A%
1980  NUMBER = RND:GOTO 2020
1990  A%=VAL(MID$(TIME$,7,2))
2000  RANDOMIZE A%
2010  NUMBER = INT(RND*(RANDOM+1))
2020  CLS:LOCATE 12,30:PRINT "Number is: ";NUMBER
2030  LOCATE 20,25:GOSUB 2860
2040  GOTO 1740
2050  REM:Subroutine for N random number
```

```
2060  CLS
2070  LOCATE 12,20:PRINT "How many random numbers to
          generate?"
2080  LOCATE 13,20:PRINT "(up to 1000)";:INPUT
          RNGNBR$
2090  IF RNGNBR$="" THEN 2050
2100  RNGNBR=VAL(RNGNBR$)
2110  IF RNGNBR < 2 THEN 2050
2120  CLS:LOCATE 12,20:PRINT "Highest number in
          range:";:INPUT RANDOM$
2130  IF RANDOM$="" THEN RANDOM$="1000"
2140  RANDOM=VAL(RANDOM$)
2150  IF RANDOM = 0 THEN 2050
2160  CLS:FOR I = 1 TO RNGNBR
2170  A%=VAL(MID$(TIME$,7,2))
2180  RANDOMIZE A%
2190  RANNUM(I)=INT(RND*(RANDOM+.1))
2200  PRINT USING "#######.";RANNUM(I);
2210  TIMELOOP=TIMER:WHILE TIMER<TIMELOOP
          +1:WEND:NEXT I
2220  LOCATE 24,25:GOSUB 2860
2230  CLS:LOCATE 12,25:PRINT "Do you want a hard copy
          print out?"
2240  LOCATE 14,36:PRINT "(Y)ES  (N)o"
2250  LOCATE 17,34:PRINT " MAKE SELECTION
          ";:YN$=INPUT$(1)
2260  YN=VAL(YN$)
2270  IF YN$="Y" THEN YN=1
2280  IF YN$="y" THEN YN=1
2290  IF YN$="N" THEN YN=2
2300  IF YN$="n" THEN YN=2
2310  IF YN < 1 THEN 2230
2320  IF YN > 2 THEN 2230
2330  IF INT(YN)=YN THEN 2340 ELSE2230
2340  ON YN GOTO 2350,1740
```

2350 REM Printout subroutine
2360 CLS:LOCATE 12,25:PRINT "Setup printer and press
       <ENTER>";:YN$=INPUT$(1)
2370 CLS:LOCATE 12,30:PRINT "PRINTING..."
2380 FOR I = 1 TO RNGNBR:LPRINT USING
       "######.";RANNUM(I):NEXT I
2390 CLS:LOCATE 12,30:PRINT "Finished Printing"
2400 TIMELOOP=TIMER:WHILE TIMER<TIMELOOP
       +1:WEND
2410 CLS:GOTO 1740
2420 GOTO 1740
2430 REM Subroutine to generator R.N. until told to stop
2440 CLS:LINE (440,140)–(130,180),3,BF:I=1
2450 LOCATE 12,20:PRINT " Highest number in range:
       ";:INPUT RNGNBR$
2460 IF RNGNBR$="" THEN RNGNBR$="1000"
2470 RNGNBR=VAL(RNGNBR$)
2480 IF RNGNBR < 2 THEN 2430
2490 CLS:LINE (400,140)–(130,180),3,BF
2500 A%=VAL(MID$(TIME$,7,2))
2510 RANDOMIZE A%
2520 RANNUM(I) = INT(RND*(RNGNBR+1))
2530 LOCATE 12,20:PRINT "Random number";I;"is:
       ";RANNUM(I)
2540 COLOR 15:LOCATE 16,24:PRINT "Press Any Key To
       Stop":COLOR 14
2550 TIMELOOP=TIMER:WHILE
       TIMER<TIMELOOP+.7:WEND
2560 I=I+1
2570 A$=INKEY$:WHILE A$="":GOTO 2490:WEND
2580 CLS:LINE (420,250)–(130,120),3,BF
2590 LOCATE 12,26:PRINT "Hard copy printout?"
2600 LOCATE 14,29:PRINT "(Y)es or N)o"
2610 LOCATE 16,27:PRINT " MAKE SELECTION
       ";:YN$=INPUT$(1)

```
2620  IF YN$="" THEN 2580
2630  IF YN$="Y" THEN 2670
2640  IF YN$="y" THEN 2670
2650  IF YN$="N" THEN 2730
2660  IF YN$="n" THEN 2730
2670  CLS:LOCATE 12,20:PRINT "Setup printer and press
         <ENTER>";:YN$=INPUT$(1)
2680  CLS:LOCATE 12,30:PRINT "Printing..."
2690  FOR II = 1 TO I–1
2700  LPRINT USING "########.#";RANNUM(II):NEXT II
2710  CLS:LOCATE 12,23:PRINT "Finished printing"
2720  TIMELOOP=TIMER:WHILE
         TIMER<TIMELOOP+1:WEND
2730  GOTO 1740
2740  CLS:RETURN
2750  XX$=INPUT$(1):RETURN
2760  REM SUBROUTINE TO ACQUIRE USER'S NAME
2770  CLS:LINE (500,200)–(125,150),3,BF
2780  LOCATE 13,20:PRINT " What is your name, please:
         ";:INPUT C$
2790  IF C$="" THEN C$="Anonymous User "
2800  CLS:LINE (500,200)–(125,150),3,BF
2810  C=LEN(C$)
2820  SPACE=32:SPACE=SPACE–(C/2)
2830  LOCATE 13,SPACE:PRINT " Thank you, ";C$
2840  TIMELOOP=TIMER:WHILE TIMER < TIMELOOP
         +1:WEND
2850  CLS:RETURN
2860  PRINT "Press any key to continue"
2870  A$=INKEY$:IF A$="" THEN 2870
2880  RETURN
2890  RETURN
2900  REM SUBROUTINE FOR KEYBOARD DATA
         ENTRY (4 August 1991)
2910  TI$=TIME$:H$=DATE$
```

```
2920  CLS:SCREEN 9:COLOR 7,1:COLOR 14
2930  LINE (490,200)–(155,120),3,BF
2940  LOCATE 12,25:PRINT " Data Collection Option
        Selected "
2950  TIMELOOP=TIMER:WHILE
        TIMER<TIMELOOP+2:WEND:CLS
2960  CLS:LINE (490,250)–(155,120),3,BF
2970  LOCATE 11,25:PRINT " Select data type:"
2980  LOCATE 13,25:PRINT " (S)ingle point (X) data   "
2990  LOCATE 14,25:PRINT " (O)rdered pair (X,Y) data "
3000  LOCATE 15,25:PRINT " (R)eturn to main menu     "
3010  LOCATE 17,25:PRINT " MAKE SELECTION
        ";:F$=INPUT$(1)
3020  CLS:IF F$="" THEN 2960:REM Discern intent of user
        on F$
3030  IF F$="S" THEN F=1
3040  IF F$="s" THEN F=1
3050  IF F$="O" THEN F=2
3060  IF F$="o" THEN F=2
3070  IF F$="R" THEN F=3
3080  IF F$="r" THEN F=3
3090  IF F < 1 THEN 2960
3100  IF F > 3 THEN 2960
3110  IF INT(F)=F THEN 3120 ELSE2960
3120  ON F GOSUB 3140,3850:REM Go to X or XY
        subroutine as selected
3130  RETURN
3140  REM Routine for X(i) data input [Ver. 1, 4 July 1991]
3150  REM Amended for STATS7 4 Augusy 1991
3160  CLS:LOCATE 12,30
3170  IF DCF=1 THEN GOTO 3780
3180  IF DCF=0      GOTO 3840
3190  N=1:NNUM=0:DC=1:DCF=1
3200  REM (Replaced code)
3210  IF ADD=1 THEN N=NNUM+1
```

```
3220 PRINT TAB(20);"X(";N;") = ";:INPUT X$
3230 IF X$="" THEN 3440 ELSE3240
3240 IF X$="" THEN 3220
3250 IF X$="9.1" THEN X$="9.100001"
3260 IF X$="-9.1" THEN X$="-9.100001"
3270 IF X$="-8.1" THEN X$="-8.100001"
3280 IF X$="8.1" THEN X$="8.100001"
3290 REM Test for minus sign
3300 XC$=LEFT$(X$,1):IF XC$ = "-" THEN Q = -1 ELSE
     Q=1
3310 IF XC$="-" THEN X$=RIGHT$(X$,LEN(X$)-1)
3320 REM Strip right-hand decimal point
3330 XC$=RIGHT$(X$,1):IF XC$="." THEN
     X$=LEFT$(X$,LEN(X$)-1)
3340 REM Strip right-hand decimal point and following zero
3350 XB$=RIGHT$(X$,2):IF XB$=".0" THEN
     X$=LEFT$(X$,LEN(X$)-2)
3360 REM Suppress leading zero for decimal fraction
3370 XC$=LEFT$(X$,1):IF XC$="0" THEN
     X$=RIGHT$(X$,LEN(X$)-1)
3380 X(N) = VAL(X$):XX$=STR$(X(N))
3390 REM Strip spurious space symbol from XX$
3400 XX$=RIGHT$(XX$,LEN(XX$)-1)
3410 IF X$=XX$ GOTO 3420 ELSE3640:REM Test for
     correct value
3420 X=Q*X:REM Add minus sign if original was negative
3430 N=N+1:GOTO 3220
3440 REM Subroutine to test for finished entering data
3450 BEEP:CLS:LINE (490,200)-(155,120),3,BF
3460 LOCATE 11,28:PRINT " Finished Entering X(i) Data?
     ":PRINT
3470 LOCATE 13,33:PRINT " (Y)es or (N)o?
     ";:YN$=INPUT$(1)
3480 IF YN$="Y" GOTO 3540:REM Determine if save to
     disk
```

```
3490 IF YN$="y" GOTO 3540:REM    "    "  "   "   "
3500 IF YN$="N" GOTO 3530:REM Go back to data
      collection routine
3510 IF YN$="n" GOTO 3530:REM "   "  "  "    "       "
3520 GOTO 3440
3530 CLS:LOCATE 12,30:GOTO 3240
3540 REM Subroutine to determine end of data collection
      orders
3550 BEEP:CLS:LINE (490,200)–(155,120),3,BF
3560 LOCATE 12,23:PRINT " Save data to disk?  (Y)es or
      (N)o? ";:YN$=INPUT$(1)
3570 IF YN$="Y" THEN 3620:REM Save data
3580 IF YN$="y" THEN 3620:REM  "   "
3590 IF YN$="N" THEN 3760:REM Go back to main
      program
3600 IF YN$="n" THEN 3760:REM "   "  "   "   "
3610 GOTO 3540
3620 N=N–1:NNUM=N:GOSUB 5030:REM Go to data file
      creation subroutine
3630 GOTO 3770
3640 BEEP:REM Error message for X$ not equal XX$
3650 CLS:LINE (490,200)–(155,120),12,BF:REM Create
      error box
3660 LOCATE 10,35:PRINT " Error! "
3670 LOCATE 12,25:PRINT " You entered ";X$;" Please
      confirm "
3680 LOCATE 13,25:PRINT " Right? (Y)es or (N)o
      ";:YN$=INPUT$(1)
3690 IF YN$="Y" THEN 3420:REM Continue collecting data
3700 IF YN$="y" THEN 3420:REM  "       "       "
3710 IF YN$="N" THEN 3740:REM Go back get new data
      point
3720 IF YN$="n" THEN 3740:REM "   "   "   "   "   "
3730 GOTO 3640
3740 PRINT:PRINT
```

```
3750 GOTO 3220
3760 NNUM=N–1
3770 RETURN
3780 BEEP:CLS:LINE (490,220)–(155,120),12,BF:REM
       Create error box
3790 LOCATE 10,36:PRINT " WARNING "
3800 LOCATE 12,27:PRINT " Data already collected      "
3810 LOCATE 13,27:PRINT " Please select another option "
3820 LOCATE 15,27:GOSUB 2860
3830 GOTO 3770
3840 GOTO 3190
3850 REM ROUTINE FOR X,Y DATA INPUT [4 August
       1991]
3860 CLS:LOCATE 12,30
3870 IF DCF=1 THEN GOTO 4500:REM Warning that data
       already collected
3880 IF DCF=0 GOTO 4560:REM Dimension X and Y (if
       needed)
3890 N=1:NNUM=0:DC=2:DCF=1:REM Set counters and
       flags.
3900 REM (Replaced code)
3910 IF ADD=1 THEN N=NNUM+1
3920 REM X-data input portion of subroutine.
3930 PRINT TAB(20);"X(";N;") = ";:INPUT X$:REM Input
       an X(i) point
3940 IF X$="" THEN 4160 ELSE3950:REM Test for finished
       collecting
3950 IF X$="" THEN 3930:REM Continue collecting (go
       back to input stmnt)
3960 IF X$="9.1" THEN X$="9.100001":REM Correct for
       rounding error
3970 IF X$="–9.1" THEN X$="–9.100001":REM "   "    "    "
3980 IF X$="–8.1" THEN X$="–8.100001":REM "   "    "    "
3990 IF X$="8.1" THEN X$="8.100001":REM   "    "    "    "
4000 REM Test for minus sign
```

```
4010 XC$=LEFT$(X$,1):IF XC$ = "–" THEN Q = –1 ELSE
        Q=1
4020 IF XC$="–" THEN X$=RIGHT$(X$,LEN(X$)–1)
4030 REM Strip right–hand decimal point
4040 XC$=RIGHT$(X$,1):IF XC$="." THEN
        X$=LEFT$(X$,LEN(X$)–1)
4050 REM Strip right–hand decimal point and following zero
4060 XB$=RIGHT$(X$,2):IF XB$=".0" THEN
        X$=LEFT$(X$,LEN(X$)–2)
4070 REM Suppress leading zero for decimal fraction
4080 XC$=LEFT$(X$,1):IF XC$="0" THEN
        X$=RIGHT$(X$,LEN(X$)–1)
4090 X(N) = VAL(X$):XX$=STR$(X(N))
4100 REM Strip spurious space symbol from XX$
4110 XX$=RIGHT$(XX$,LEN(XX$)–1)
4120 IF X$=XX$ GOTO 4130 ELSE4360:REM Test for
        correct value
4130 X=Q*X:REM Add minus sign if original was negative
4140 GOTO 4570:REM Go get a Y-data point
4150 N=N+1:GOTO 3930
4160 REM Subroutine to test for finished entering data
4170 CLS:LINE (490,200)–(155,120),3,BF
4180 LOCATE 12,29:PRINT " Finished Entering Data?
        ":PRINT
4190 LOCATE 13,29:PRINT " (Y)es or (N)o?
        ";:YN$=INPUT$(1)
4200 IF YN$="Y" GOTO 4260:REM Determine if save to disk
4210 IF YN$="y" GOTO 4260:REM    "    "  "   "   "
4220 IF YN$="N" GOTO 4250:REM Go back to data
        collection routine
4230 IF YN$="n" GOTO 4250:REM "  "  "  "   "       "
4240 GOTO 4160
4250 CLS:LOCATE 12,30:GOTO 3950
4260 REM Subroutine to determine end of data collection
        orders
```

```
4270  CLS:LINE (490,200)–(155,120),3,BF
4280  LOCATE 12,24:PRINT " Save data to disk?  (Y)es or
        (N)o? ";:YN$=INPUT$(1)
4290  IF YN$="Y" THEN 4340:REM Save data
4300  IF YN$="y" THEN 4340:REM   "   "
4310  IF YN$="N" THEN 4480:REM Go back to main program
4320  IF YN$="n" THEN 4480:REM  "   "   "   "   "
4330  GOTO 4260
4340  N=N–1:NNUM=N:GOSUB 5030:REM Go to data file
        creation subroutine
4350  GOTO 4490
4360  BEEP:REM Error message for X$ not equal XX$
4370  CLS:LINE (490,200)–(155,120),12,BF:REM Create
        error box
4380  LOCATE 10,35:PRINT " Error! "
4390  LOCATE 12,25:PRINT "You entered";X$;"Please confirm"
4400  LOCATE 13,25:PRINT " Right? (Y)es  or  (N)o
        ";:YN$=INPUT$(1)
4410  IF YN$="Y" THEN 4130:REM Continue collecting data
4420  IF YN$="y" THEN 4130:REM   "       "       "
4430  IF YN$="N" THEN 4460:REM Go back get new data
        point
4440  IF YN$="n" THEN 4460:REM "   "   "   "   "   "
4450  GOTO 4360
4460  PRINT:PRINT
4470  GOTO 3930
4480  NNUM=N–1
4490  RETURN
4500  BEEP:CLS:LINE (490,220)–(155,120),12,BF:REM
        Create error box
4510  LOCATE 10,36:PRINT " WARNING "
4520  LOCATE 12,27:PRINT " Data already collected      "
4530  LOCATE 13,27:PRINT " Please select another option "
4540  LOCATE 15,27:GOSUB 2860
4550  GOTO 4490
```

```
4560 GOTO 3890
4570 REM ROUTINE FOR Y(i) DATA INPUT [Ver. 1, 8
     July 1991]
4580 REM Y-data input portion of subroutine.
4590 PRINT TAB(20);"Y(";N;") = ";:INPUT Y$:REM Input
     an Y(i) point
4600 IF Y$="" THEN 4910 ELSE4610:REM Test for finished
     collecting
4610 IF Y$="" THEN 4590:REM Continue collecting (go
     back to input stmnt)
4620 IF Y$="9.1" THEN Y$="9.100001":REM Correct for
     rounding error
4630 IF Y$="-9.1" THEN Y$="-9.100001":REM "   "   "   "
4640 IF Y$="-8.1" THEN Y$="-8.100001":REM "   "   "   "
4650 IF Y$="8.1" THEN Y$="8.100001":REM   "   "   "   "
4660 REM Test for minus sign
4670 YC$=LEFT$(Y$,1):IF YC$ = "-" THEN Q = -1 ELSE
     Q=1
4680 IF YC$="-" THEN Y$=RIGHT$(Y$,LEN(Y$)-1)
4690 REM Strip right-hand decimal point
4700 YC$=RIGHT$(Y$,1):IF YC$="." THEN
     Y$=LEFT$(Y$,LEN(Y$)-1)
4710 REM Strip right-hand decimal point and following zero
4720 YB$=RIGHT$(Y$,2):IF YB$=".0" THEN
     Y$=LEFT$(Y$,LEN(Y$)-2)
4730 REM Suppress leading zero for decimal fraction
4740 YC$=LEFT$(Y$,1):IF YC$="0" THEN
     Y$=RIGHT$(Y$,LEN(Y$)-1)
4750 Y(N) = VAL(Y$):YY$=STR$(Y(N))
4760 REM Strip spurious space symbol from YY$
4770 YY$=RIGHT$(YY$,LEN(YY$)-1)
4780 IF Y$=YY$ GOTO 4790 ELSE4810:REM Test for
     correct value
4790 Y=P*Y:REM Add minus sign if original was negative
```

```
4800 GOTO 4150:REM Return to X-Y data collection at X-
       input point
4810 BEEP:REM Error message for Y$ not equal YY$
4820 CLS:LINE (490,200)–(155,120),12,BF:REM Create
       error box
4830 LOCATE 10,35:PRINT " Error! "
4840 LOCATE 12,25:PRINT "You entered";X$;"Please
       confirm"
4850 LOCATE 13,25:PRINT " Right?  (Y)es  or  (N)o
       ";:YN$=INPUT$(1)
4860 IF YN$="Y" THEN 4790:REM Continue collecting data
       points
4870 IF YN$="y" THEN 4790:REM     "       "        "      "
4880 IF YN$="N" THEN 4900:REM Go back and get new
       data point
4890 IF YN$="n" THEN 4900:REM "  "  "  "  "  "     "
4900 GOTO 4590
4910 REM Test for finished collecting data
4920 CLS:LINE (500,220)–(125,115),12,BF
4930 LOCATE 10,35:PRINT " WARNING! "
4940 LOCATE 12,20:PRINT " Unequal number of X and Y
       data points. "
4950 LOCATE 13,20:PRINT " (Missing Y will be set to zero) "
4960 LOCATE 15,30:PRINT " OK?  (Y)es or (N)o
       ";:YN$=INPUT$(1)
4970 IF YN$="Y" THEN GOTO 4260:REM Determine if
       save to disk
4980 IF YN$="y" THEN GOTO 4260:REM     "      "  "  "  "
4990 IF YN$="N" GOTO 4580:REM Go back to data
       collection routine
5000 IF YN$="n" GOTO 4580:REM "  "  "  "    "      "
5010 GOTO 4910
5020 CLS:LOCATE 12,30:GOTO 4610
5030 REM DUMP DATA TO DISK SUBROUTINE
5040 GOSUB 5050:RETURN
```

```
5050  REM SUBROUTINE TO DUMP DATA TO
      DISKETTE
5060  A$="X:AAAAAAAA":REM Dummy variable for
      diskette file
5070  CLS:SCREEN 9:COLOR 7,1:COLOR 14:REM Get disk
      drive no.
5080  LINE (500,210)–(135,140),3,BF
5090  LOCATE 12,27:PRINT " Drive for data disk     "
5100  LOCATE 13,27:PRINT " Format is a: b: c: or d: "
5110  LOCATE 14,27:PRINT " SELECTION: ";:INPUT B$
5120  IF B$="" THEN BEEP:CLS:IF B$="" THEN 5080
5130  B$=LEFT$(B$,1)
5140  IF B$="a" THEN 5230
5150  IF B$="A" THEN 5230
5160  IF B$="b" THEN 5230
5170  IF B$="B" THEN 5230
5180  IF B$="c" THEN 5230
5190  IF B$="C" THEN 5230
5200  IF B$="d" THEN 5230
5210  IF B$="D" THEN 5230
5220  BEEP:CLS:GOTO 5080
5230  MID$(A$,1)=B$:REM Change disk drive to selected
5240  CLS:LINE (550,250)–(80,120),3,BF
5250  LOCATE 12,15:PRINT " ENTER NAME OF DATA
      FILE "
5260  LOCATE 14,15:PRINT " Limit EIGHT alphabet
      characters. Program will   "
5270  LOCATE 15,15:PRINT " truncate filenames longer
      than EIGHT characters."
5280  LOCATE 17,15:PRINT " Filename is: ";:INPUT D$
5290  IF D$="" THEN BEEP:IF D$="" THEN 5240
5300  CLS:D$=LEFT$(D$,8)
5310  LINE (500,200)–(125,150),3,BF
5320  LOCATE 13,27:PRINT "Writing data to file ";B$;":";D$;""
5330  MID$(A$,3)=D$:A$=LEFT$(A$,LEN(D$)+2)
```

```
5340  OPEN "O",#1,A$:PRINT#1,NNUM;DC;D$;",";C$;",";
         H$;",";TI$:CLOSE#1
5350  OPEN "A",#1,A$
5360  FOR I = 1 TO N:PRINT#1,X(I),Y(I):NEXT I
5370  CLOSE #1:RETURN
5380  REM SUBROUTINE TO RETRIEVE X(i) DATA
         FROM DISK
5390  IF DCF=1 THEN GOTO 5870
5400  GA$="Incorrect disk drive (A through D legal)"
5410  GB$="Illegal Filename":DCF=1
5420  CLS:SCREEN 9:LINE (500,175)–(100,120),3,BF
5430  LOCATE 11,15:PRINT " Retrieve file [(drive):name]:
         ";:INPUT G$
5440  IF G$="" THEN 5420 ELSE5450
5450  GC$=LEFT$(G$,2):IF LEN(G$)<2 THEN 5420
         ELSE5460
5460  B$=LEFT$(G$,2):H$=RIGHT$(G$,LEN(G$)–2)
5470  IF B$="A:" THEN 5560
5480  IF B$="a:" THEN 5560
5490  IF B$="B:" THEN 5560
5500  IF B$="b:" THEN 5560
5510  IF B$="C:" THEN 5560
5520  IF B$="c:" THEN 5560
5530  IF B$="D:" THEN 5560
5540  IF B$="d:" THEN 5560
5550  PRINT:PRINT TAB(20);GA$:GOSUB 2860:GOTO
         5420
5560  IF LEN(H$) < 1 THEN BEEP:PRINT
         TAB(20);GB$:GOSUB 2860
5570  IF LEN(H$) < 1 THEN 5420 ELSE5580
5580  IF LEN(H$) > 8 THEN BEEP:PRINT
         TAB(20);GB$:PRINT:GOSUB 2860
5590  IF LEN(H$) > 8 THEN 5420 ELSE5600
5600  CLS:LINE (450,200)–(125,150),3,BF
5610  LOCATE 13,26:PRINT " Retrieving file ";G$;" "
```

```
5620  OPEN "I",#1,G$
5630  INPUT #1,NNUM,DC,D$,C$,H$,TI$
5640  FOR I = 1 TO NNUM:INPUT#1,X(I),Y(I):NEXT I
5650  CLOSE#1
5660  CLS:LINE (450,240)–(125,130),3,BF
5670  LOCATE 12,20:PRINT " Number of data X(i) points:
         ";NNUM;" "
5680  LOCATE 13,20:PRINT " DC = ";DC;" "
5690  LOCATE 14,20:PRINT " Retrieved file is: ";D$;" "
5700  LOCATE 15,20:PRINT " Original file recorded by:
         ";C$;" "
5710  LOCATE 16,20:PRINT "Date file data recorded:";H$;" "
5720  LOCATE 20,20:GOSUB 2860
5730  CLS:LINE (450,200)–(125,120),3,BF
5740  LOCATE 11,30:PRINT " Review data? "
5750  LOCATE 13,30:PRINT " (Y)es or (N)o
         ";:YN$=INPUT$(1)
5760  IF YN$="N" THEN 5920
5770  IF YN$="n" THEN 5920
5780  IF YN$="Y" THEN 5810
5790  IF YN$="y" THEN 5810
5800  GOTO 5730
5810  CLS
5820  LOCATE 2,20:PRINT TAB(20);" Data are:
         ":PRINT:PRINT
5830  FOR I = 1 TO NNUM
5840  PRINT TAB(20);"X(";I;") = ";X(I);"   Y(";I;") =
         ";Y(I):NEXT I
5850  PRINT:PRINT TAB(20):GOSUB 2860:RETURN
5860  RETURN
5870  CLS:BEEP:LINE (400,225)–(170,150),12,BF
5880  LOCATE 12,32:PRINT " WARNING "
5890  LOCATE 13,25:PRINT " Data already collected "
5900  LOCATE 14,25:PRINT " Select another option. "
5910  LOCATE 16,25:GOSUB 2860
```

```
5920 RETURN
5930 REM SUBROUTINE TO PERFORM STATISTICAL
     CALCULATIONS (CALCULATE)
5940 IF DCF=0 THEN 6520:REM Go to error msg. if no data
     collected
5950 IF CALCDONE = 1 THEN 6470
5960 DIM MINX(1000):DIM MINY(1000)
5970 CLS:LINE (500,220)–(125,100),3,BF
5980 LOCATE 9,34:PRINT "          "
5990 LOCATE 10,34:PRINT " DOING MATH "
6000 LOCATE 11,34:PRINT "          "
6010 LOCATE 12,26:PRINT " Please wait...and be patient "
6020 LOCATE 13,20:PRINT "                              "
6030 LOCATE 14,20:PRINT " I may be slow, but I'm faster
     than you! "
6040 REM Subroutine to sort X,Y to find minimum &
     maximum values
6050 FOR I = 1 TO NNUM: REM Preset X & Y arrays
6060 MINX(I) – X(I)
6070 MINY(I) = Y(I)
6080 NEXT I
6090 FOR JJ = 1 TO NNUM – 1
6100 FOR KK = JJ + 1 TO NNUM
6110 IF MINX(KK) < MINX(JJ) THEN SWAP
     MINX(KK),MINX(JJ)
6120 IF MINY(KK) < MINY(JJ) THEN SWAP
     MINY(KK),MINY(JJ)
6130 NEXT KK
6140 NEXT JJ
6150 XMIN=MINX(1):XMAX=MINX(NNUM):YMIN=
     MINY(1):YMAX=MINY(NNUM)
6160 REM Calculate values for statistics
6170 FOR I = 1 TO NNUM
6180 SUMX = SUMX+X(I):SUMY=SUMY+Y(I):REM
     Calculate sums of X & Y
```

6190 REM Calculate the sum of X-squared and Y-squared
6200 SUMXX = SUMXX + (X(I)^2):SUMYY=SUMYY+
     (Y(I)^2)
6210 SUMXY=SUMXY + (X(I)*Y(I)):REM Calc. sum of XY
     product
6220 NEXT I
6230 XBAR = SUMX/NNUM:YBAR = SUMY/NNUM
6240 SUMMX=SUMX^2:SUMMY=SUMY^2:REM Calc.
     square of sums of X and Y
6250 NUMERX = SUMXX – (SUMMX/NNUM):NUMERY
     = SUMYY – (SUMMY/NNUM)
6260 VARXS=NUMERX/(NNUM–1):VARXP=NUMERX/
     (NNUM):REM Calc. Var. of X
6270 VARYS=NUMERY/(NNUM–1):VARYP=NUMERY/
     (NNUM):REM Calc. Var. of Y
6280 SDEVXS=SQR(VARXS):SDEVXP=SQR(VARXP):
     REM Calc. Std. Dev. of X
6290 SDEVYS=SQR(VARYS):SDEVYP=SQR(VARYP):
     REM Calc. Std. Dev. of Y
6300 REM Calculate "a" and "b" in linear regression
6310 IF DC=1 THEN 6560:REM Bypass if X (not X,Y) data is
     present
6320 NUMERLR=SUMXY–((SUMX*SUMY))/NNUM:
     REM Conventional Linear Regression
6330 DENOM = SUMXX – (SUMMX/NNUM)
6340 CB = NUMERLR/DENOM
6350 CA = (SUMY/NNUM) – ((CB*SUMX)/NNUM)
6360 CR = NUMERLR/((NNUM–1)*SQR(VARXS)
     *SQR(VARYS))
6370 REM Orthogonal linear regression
6380 SXX = SUMXX – (SUMMX/NNUM):SYY=SUMYY –
     (SUMMY/NNUM)
6390 SXY= SUMXY – ((SUMX*SUMY)/NNUM)
6400 SD = SYY–SXX
6410 NUMEROR = SD + SQR(SD^2 + (4*SXY^2))

```
6420 OB = NUMEROR/(2*SXY)
6430 OA = (SUMY/NNUM) – (OB*SUMX)/NNUM
6440 SB = SXX*SYY:SB = SQR(SB)
6450 RO = SXY/SB
6460 GOTO 6560
6470 REM Subroutine for calculation done error message
6480 BEEP:CLS:LOCATE 12,30:PRINT " ERROR! "
6490 LOCATE 13,30:PRINT " Calculations already done.
     Select another option "
6500 LOCATE 14,30:PRINT " from main menu.
     ":PRINT:GOSUB 2900
6510 RETURN
6520 REM Subroutine for calculations when data not collected
6530 CLS:BEEP:LOCATE 12,30:PRINT "ERROR!":PRINT
6540 COLOR 14:PRINT TAB(30);"No data collected yet.
6550 LOCATE 16,30:GOSUB 2860:CLS:RETURN
6560 GOSUB 6580:REM Subroutine to print results to screen
6570 CALCDONE=1:RETURN
6580 REM SCREEN PRINTOUT SUBROUTINE
6590 CLS:SCREEN 9:LINE (450,175)–(125,140),3,BF
6600 LOCATE 12,31:PRINT " Printing... "
6610 TIMELOOP=TIMER:WHILE TIMER<TIMELOOP
     +1:WEND
6620 IF D$="" THEN D$="(no file name)"
6630 CLS:LINE (550,250)–(125,100),3,BF
6640 LOCATE 10,20:PRINT " Printout of file: ";D$;" "
6650 LOCATE 11,20:PRINT " and calculated results      "
6660 LOCATE 13,20:PRINT " File recorded by: ";C$;" "
6670 LOCATE 14,20:PRINT " Date: ";H$;" Time: ";TI$;" "
6680 LOCATE 15,20:PRINT " Number of Data points:
     ";NNUM;" "
6690 LOCATE 17,20:GOSUB 2860
6700 CLS:LOCATE 1,10:PRINT " Data are: ":PRINT
6710 IF DC=1 THEN 6730:REM Printout X(i) data
6720 IF DC=2 THEN 6750:REM Printout [X,Y] data
```

```
6730 FOR I = 1 TO NNUM:PRINT "X(";I;") = ";X(I);"
     ":NEXT I
6740 GOTO 6770
6750 FOR I=1 TO NNUM:PRINT "X(";I;") = ";X(I);"
     Y(";I;") = ";Y(I)
6760 NEXT I:PRINT:PRINT
6770 PRINT TAB(20):GOSUB 2860:CLS
6780 PRINT:PRINT TAB(5);"Statistics for X-data":PRINT
6790 PRINT TAB(5);"Min. Val.: ";XMIN;" Max. Val.:
     ";XMAX
6800 PRINT TAB(5);"Range: ";XMIN;" to ";XMAX
6810 PRINT TAB(5);"Average (mean): ";XBAR
6820 PRINT TAB(5);"Variance (sample): ";VARXS
6830 PRINT TAB(5);"Variance (Population): ";VARXP
6840 PRINT TAB(5);"Standard deviation (sample):
     ";SDEVXS
6850 PRINT TAB(5);"Standard deviation (population):
     ";SDEVXP
6860 IF DC=2 THEN 6870 ELSE7050
6870 REM Subroutine if Y-data present
6880 PRINT:PRINT:PRINT TAB(5);"Statistics for Y-
     data":PRINT
6890 PRINT TAB(5);"Min. Val.: ";YMIN;" Max. Val.:
     ";YMAX
6900 PRINT TAB(5);"Range: ";YMIN;" to ";YMAX
6910 PRINT TAB(5);"Average (mean): ";YBAR
6920 PRINT TAB(5);"Variance (sample): ";VARYS
6930 PRINT TAB(5);"Variance (Population): ";VARYP
6940 PRINT TAB(5);"Standard deviation (sample):
     ";SDEVYS
6950 PRINT TAB(5);"Standard deviation (population):
     ";SDEVYP
6960 PRINT:GOSUB 2860:CLS:PRINT:PRINT
     TAB(5);"Linear regression results":PRINT
```

6970 PRINT:PRINT TAB(5);"For Conventional Least
     Squares":PRINT
6980 PRINT TAB(5);"Slope, b = ";CB
6990 PRINT TAB(5);"Y–intercept, a = ";CA
7000 PRINT TAB(5);"Correlation, r = ";CR
7010 PRINT:PRINT:PRINT TAB(5);"For Orthogonal least
     squares":PRINT
7020 PRINT TAB(5);"Slope, b = ";OB
7030 PRINT TAB(5);"Y–intercept, a = ";OA
7040 PRINT TAB(5);"Correlation, r = ";RO
7050 PRINT:GOSUB 2860
7060 RETURN
7070 REM HARDCOPY PRINTOUT SUBROUTINE
7080 CLS:LINE (450,200)–(125,120),3,BF
7090 LOCATE 12,30:PRINT " Printing... "
7100 IF D$="" THEN D$="(no file name)"
7110 LPRINT TAB(5);"Printout of file ";D$;" and calculated
     results"
7120 LPRINT:LPRINT TAB(7);"File recorded by: ";C$
7130 LPRINT TAB(7);"Date: ";H$;" Time: ";TI$
7140 LPRINT TAB(7);"Number of Data points: ";NNUM:
     LPRINT
7150 LPRINT TAB(7);"Data are: ":LPRINT
7160 IF DC=1 THEN 7180:REM Printout X(i) data
7170 IF DC=2 THEN 7200:REM Printout [X,Y] data
7180 FOR I = 1 TO NNUM:LPRINT TAB(7); "X(";I;") =
     ";X(I):NEXT I
7190 GOTO 7220
7200 FOR I=1 TO NNUM:LPRINT TAB(7);"X(";I;") =
     ";X(I);"   Y(";I;") = ";Y(I)
7210 NEXT I:LPRINT:LPRINT
7220 LPRINT:LPRINT TAB(5);"Statistics for X-data":
     LPRINT
7230 LPRINT TAB(5);"Min. Val.: ";XMIN;" Max. Val.:
     ";XMAX

```
7240 LPRINT TAB(5);"Range: ";XMIN;" to ";XMAX
7250 LPRINT TAB(5);"Average (mean): ";XBAR
7260 LPRINT TAB(5);"Variance (sample): ";VARXS
7270 LPRINT TAB(5);"Variance (Population): ";VARXP
7280 LPRINT TAB(5);"Standard deviation (sample):
      ";SDEVXS
7290 LPRINT TAB(5);"Standard deviation (population):
      ";SDEVXP
7300 IF DC=2 THEN 7310 ELSE7490
7310 REM Subroutine if Y-data present
7320 LPRINT:LPRINT TAB(5);"Statistics for Y-
      data":LPRINT
7330 LPRINT TAB(5);"Min. Val.: ";YMIN;" Max. Val.:
      ";YMAX
7340 LPRINT TAB(5);"Range: ";YMIN;" to ";YMAX
7350 LPRINT TAB(5);"Average (mean): ";YBAR
7360 LPRINT TAB(5);"Variance (sample): ";VARYS
7370 LPRINT TAB(5);"Variance (Population): ";VARYP
7380 LPRINT TAB(5);"Standard deviation (sample):
      ";SDEVYS
7390 LPRINT TAB(5);"Standard deviation (population):
      ";SDEVYP
7400 LPRINT:LPRINT TAB(5);"Linear regression
      results":LPRINT
7410 LPRINT TAB(5);"For Conventional Least Squares"
7420 LPRINT TAB(5);"Slope, b = ";CB
7430 LPRINT TAB(5);"Y-intercept, a = ";CA
7440 LPRINT TAB(5);"Correlation, r = ";CR
7450 LPRINT:LPRINT TAB(5);"For Orthogonal least
      squares":LPRINT
7460 LPRINT TAB(5);"Slope, b = ";OB
7470 LPRINT TAB(5);"Y-intercept, a = ";OA
7480 LPRINT TAB(5);"Correlation, r = ";RO
7490 CLS:LOCATE 12,20:GOSUB 2860
7500 RETURN
```

APPENDIX C

OVENMASTER II

Ovenmaster II is a computer simulation game, written in MS-DOS BASIC (i.e., BASICA or GWBASIC), that is designed to teach a lesson in experimentation. The object is to determine the correct oven temperature required to heat cure a batch of epoxy adhesives in a laboratory or factory. The player selects a temperature and a curing time, and then commands the "oven" to do its job. When the oven is finished, a judgment is made on the quality of the adhesive bond.

The idea behind the game is to learn to collect adequate data to justify your conclusion. Taking a single data point is not the right answer! Some "real worldness" is added to the problem because there are six ovens: two are normal and in control, two are normal but border on being out of control, and two are completely out of control (producing random results). Which is which? Play the game and find out!

Executable diskette available from the author, P.O. Box 1099, Falls Church, VA 22041-0099.

```
10    REM Program Name: OVENMASTER II Ver. 1.1
         (OVEN6) [30 July 1991]
20    CLEAR: KEY OFF:SCREEN 0
30    REM TABLE OF CONSTANTS AND DIMENSION
         STATEMENTS
```

```
40    KEY OFF
50    B$="Please Make Selection"
60    G$ = " (Times Different Each Trial)"
70    H$ = " (Temperature Different Each Trial)"
80    DIM C$(100)
90    DIM TT(100)
100   NZ=0
110   DIM AT(1000)
120   DIM DA(1000)
130   DIM AX(1000)
140   DIM SB(1000)
150   DIM MINT(1000)
160   DIM MIND(1000)
170   REM DICTIONARY OF VARIABLES USED IN THIS
         PROGRAM
180   REM D = Doneness
190   REM T = Time for curing (hours)
200   REM ST = Set temperature (500 to 1000 degrees)
210   REM AT = Actual temperature (ST with some
         randomness added)
220   REM AX = Dummy variable for printing temperature
230   REM DA = Dummy variable for printing doneness
240   REM NT = Number of trials in multiple trials protocols
250   REM TBAR = Mean temperature all trials
260   REM DBAR = Mean doneness all trials
270   REM VARD = Doneness variance
280   REM VART = Temperature variance
290   REM DEVD = Doneness standard deviation
300   REM DEVT = Temperature standard deviation
310   REM MINTEMP = Minimum temperature all runs
320   REM MAXTEMP = Maximum temperature all runs
330   REM MINDONE = Minimum doneness all runs
340   REM MAXDONE = Maximum doneness all runs
350   REM MINT = Dummy variable for temperature sort
360   REM MIND = Dummy variable for doneness sort
```

```
370   REM A$ = "Press Any Key..." message
380   REM C$ = Program user's name
390   REM B$ = "Make Selection..." message
400   REM K = Oven selector
410   REM PT = Multiple trials protocol selector
420   REM L = Type of trial selector
430   REM PAU = Pause/No-Pause selector
440   REM LNPRT = Printer output selector
450   REM G$ = "Times Different..." message
460   REM H$ = "Temperature Different..." message
470   REM EXECUTION SEGMENT OF PROGRAM
480   GOSUB 1110: REM Opening Screen
490   GOSUB 1220: REM Opening Announcement
500   GOSUB 1770: REM Obtain operator name (Variable =
         C$)
510   GOSUB 1490: REM Select Type of Trial (Variable = L)
520   IF L = 3 GOTO 920
530   ON L GOSUB 560,820,1710,20
540   IF L = 1 THEN GOTO 510
550   GOTO 510
560   REM SUBROUTINE FOR SINGLE TRIALS
570   CLS
580   LOCATE 12,25:PRINT "SINGLE TRIAL METHOD
         SELECTED"
590   TIMELOOP = TIMER
600   WHILE TIMER < TIMELOOP + 2:WEND
610   GOSUB 1330: REM Select Oven (K)
620   GOSUB 1880: REM Select Desired Temperature (ST)
630   GOSUB 1960: REM Select Curing Time (T)
640   ON K GOSUB 2690,2780,2690,2780,2870,2870:REM
         Get AT
650   D=(AT/1000)*T
660   D=D*100
670   D=INT(D)
680   D=D/100
```

```
690   GOSUB 2050: REM Timer To Reseed Random Number
        Generator
700   GOSUB 2350
710   CLS:GOSUB 1040
720   GOSUB 5460
730   LOCATE 10,20:PRINT "RESULTS OF THIS RUN:"
740   LOCATE 12,20:PRINT "Set Oven Temperature: ";ST;"
        Degrees F."
750   LOCATE 13,20:PRINT "Actual Oven Temperature:
        ";AT;" Degrees F."
760   LOCATE 14,20:PRINT "Curing Time: ";T;" Hours"
770   LOCATE 15,20:PRINT "Doneness Scale Result: ";D;"
        Zlotz Units"
780   LOCATE 16,20:PRINT "Subjective Opinion of
        Doneness: ";D$
790   LOCATE 17,20:PRINT D$
800   LOCATE 19,20:GOSUB 2570: REM Press Any...
810   RETURN
820   REM SUBROUTINE FOR MULTIPLE TRIALS (L=2)
830   CLS:SCREEN 9:COLOR 7,1:COLOR 14
840   LOCATE 12,25:PRINT "MULTIPLE TRIALS
        METHOD SELECTED"
850   TIMELOOP = TIMER
860   WHILE TIMER < TIMELOOP + 2:WEND
870   GOSUB 2930: REM Select Protocol (Variable = PT)
880   IF PT = 4 GOTO 510
890   GOSUB 3070: REM Select Number of Trials (Variable
        = NT)
900   ON PT GOSUB 3170,3400,3690,510: REM Go execute
        Selected Protocol
910   RETURN
920   GOSUB 1710
930   END
940   REM MINOR TEXT BOX SUBROUTINE
950   CLS
```

```
960   LINE (460,130)–(150,200),2,B
970   PAINT (200,150),3,2
980   COLOR 14:LOCATE 22,1:RETURN
990   REM MAJOR TEXT BOX SUBROUTINE
1000  CLS
1010  LINE (480,75)–(150,276),2,B
1020  PAINT (200,150),3,2
1030  COLOR 14:LOCATE 22,1:RETURN
1040  REM MAJOR WIDE TEXT BOX SUBROUTINE
1050  CLS
1060  LINE (550,75)–(75,276),2,B
1070  PAINT (200,150),3,2
1080  COLOR 14:LOCATE 22,1:RETURN
1090  REM JUMP 10 LINES SUBROUTINE
1100  LOCATE 12:RETURN
1110  REM OPENING SCREEN SUBROUTINE
1120  NTE(1)=523.25:NTE(2)=493.88:NTE(3)=523.25:
        NTE(4)=587.33:NTE(5)=659.26
1130  NTE(6)=698.46:NTE(7)=783.99:NTE(8)=880:
        NTE(9)=987.77:NTE(10)=1046.5
1140  CLS:SCREEN 9:XXX1=400:XXX2=100:YYY1=
        50:YYY2=200:M=10:COLOR 15
1150  LINE (XXX1,YYY1)–(XXX2,YYY2),,B:SOUND
        NTE(M),10
1160  M=M–1:IF M = 0 THEN 1180 ELSE1170
1170  XXX1=XXX1+10:XXX2=XXX2+10:YYY1=YYY1+10:
        YYY2=YYY2+10:GOTO 1150
1180  COLOR 14:LOCATE 12,32:PRINT "OVENMASTER
        II":COLOR 15
1190  LOCATE 14,26:PRINT "Copyright 1991 J.J. Carr"
1200  TIMELOOP=TIMER:WHILE TIMER < TIMELOOP +
        2:WEND
1210  COLOR 7,1:COLOR 14:LOCATE 22,1:RETURN
1220  REM OPENING ANNOUNCEMENT SUBROUTINE
1230  CLS:GOSUB 1040
```

```
1240  LOCATE 10,15:PRINT "This program is a simulation
        game in which you     "
1250  LOCATE 11,15:PRINT "select an adhesive curing oven
        for some tests.     "
1260  LOCATE 12,15:PRINT "The idea is to find the best
        'doneness' level to   "
1270  LOCATE 13,15:PRINT "satisfy customer needs.  On a
        scale of 1 to 10, any"
1280  LOCATE 14,15:PRINT "product between 4 and 6 is
        acceptable.  You select "
1290  LOCATE 15,15:PRINT "both the curing time and the
        oven temperature.     "
1300  LOCATE 16,15:PRINT "There are six ovens in various
        states of repair.   "
1310  LOCATE 19,28:GOSUB 2570
1320  RETURN
1330  REM SUBROUTINE TO SELECT OVEN (Variable = K)
1340  GOSUB 990
1350  LOCATE 8,30:PRINT C$;","
1360  LOCATE 9,30:PRINT "Please select oven"
1370  LOCATE 11,30:PRINT "1. Oven Number 1"
1380  LOCATE 12,30:PRINT "2. Oven Number 2"
1390  LOCATE 13,30:PRINT "3. Oven Number 3"
1400  LOCATE 14,30:PRINT "4. Oven Number 4"
1410  LOCATE 15,30:PRINT "5. Oven Number 5"
1420  LOCATE 16,30:PRINT "6. Oven Number 6"
1430  LOCATE 18,30:PRINT B$;:K$=INPUT$(1)
1440  K = VAL(K$)
1450  IF K < 1 THEN GOTO 1330
1460  IF K > 6 THEN GOTO 1330
1470  IF INT(K)=K THEN GOTO 1480 ELSE1330
1480  CLS:RETURN
1490  REM MAIN MENU TO SELECT FUNCTIONS
        (RETURNS L$,L,LX)
1500  GOSUB 990
```

```
1510  LOCATE 10,26:PRINT "Select the desired function   "
1520  LOCATE 12,26:PRINT "(S)ingle trial on one oven   "
1530  LOCATE 13,26:PRINT "(M)ultiple trials on one oven"
1540  LOCATE 14,26:PRINT "(E)nd game                   "
1550  LOCATE 15,26:PRINT "(R)estart program            "
1560  LOCATE 17,26:PRINT "Please make selection now:
         ";:L$=INPUT$(1)
1570  IF L$="0" THEN 1490
1580  IF L$ = "" THEN 1490
1590  LX = VAL(L$)
1600  IF LX > 0 THEN 1490
1610  IF L$="S" THEN L=1:IF L$="S" GOTO 1700
1620  IF L$="s" THEN L=1:IF L$="s" GOTO 1700
1630  IF L$="M" THEN L=2:IF L$="M" GOTO 1700
1640  IF L$="m" THEN L=2:IF L$="m" GOTO 1700
1650  IF L$="E" THEN L=3:IF L$="E" GOTO 1700
1660  IF L$="e" THEN L=3:IF L$="e" GOTO 1700
1670  IF L$="R" THEN L=4:IF L$="R" GOTO 1700
1680  IF L$="r" THEN L=4:IF L$="r" GOTO 1700
1690  GOTO 1490
1700  RETURN
1710  REM END PROGRAM SUBROUTINE
1720  GOSUB 940
1730  LOCATE 12,POYNT-4:PRINT "Goodbye ";C$
1740  TIMELOOP = TIMER:WHILE TIMER < TIMELOOP
         + 2:WEND
1750  SCREEN 0:CLS:KEY ON:COLOR 0,0:COLOR 7
1760  RETURN
1770  REM SUBROUTINE TO INPUT OPERATOR'S
         NAME (RETURNS C$)
1780  GOSUB 940
1790  LOCATE 12,26:PRINT "What is your name,
         please?":LOCATE 14,28:INPUT C$
1800  IF C$="" GOTO 1860 ELSE1810
1810  GOSUB 940
```

```
1820  POYNT = 20   + (19 - (LEN(C$)/2))
1830  LOCATE 12,35:PRINT "Thank You"
1840  LOCATE 14,POYNT:PRINT C$
1850  TIMELOOP=TIMER:WHILE TIMER <
      TIMELOOP+2:WEND:GOTO 1870
1860  C$="Anonymous User":GOTO 1810
1870  CLS:RETURN
1880  REM SET OVEN TEMPERATURE
1890  CLS:LINE (520,130)-(130,250),2,B:PAINT
      (200,150),3,2:COLOR 14
1900  LOCATE 13,22:PRINT C$;", Select Oven Temperature"
1910  LOCATE 15,22:PRINT "Permissable range is 500 to
      1000  degrees"
1920  LOCATE 17,22:PRINT "Desired Temperature
      Is:";:INPUT ST
1930  IF ST > 1000 THEN GOTO 1900
1940  IF ST < 500 THEN GOTO 1900
1950  CLS:RETURN
1960  REM SUBROUTINE TO SELECT CURING TIME
1970  CLS:GOSUB 940:LOCATE 11,22:PRINT C$;","
1980  LOCATE 12,22:PRINT "Please Oven Select Curing
      Time."
1990  LOCATE 13,22:PRINT "Permitted range is 1 to 10
      Hours."
2000  LOCATE 14,22:PRINT "Curing Time is: ";:INPUT T
2010  IF T < 1 THEN GOTO 1960
2020  IF T > 10 THEN GOTO 1960
2030  CLS:RETURN
2040  RETURN
2050  REM TIMER FOR ADHESIVE COOKING (Reseed
      Random No. Generator)
2060  CLS:GOSUB 940
2070  IF L=2 GOTO 2160
2080  COLOR 15:LOCATE 12,30:PRINT " ADHESIVE IS
      COOKING ":COLOR 14
```

```
2090 LOCATE 13,30:PRINT " Please Be Patient":PRINT
2100 LOCATE 14,30:PRINT "Curing Time:";T;"hours"
2110 GOSUB 5620
2120 TIMELOOP=TIMER:WHILE TIMER < TIMELOOP +
     (T/4):WEND
2130 CLS:GOSUB 940:LOCATE 12,33:PRINT "BATCH IS
     DONE"
2140 TIMELOOP = TIMER:WHILE TIMER < TIMELOOP
     + 2.5:WEND
2150 COLOR 7,1:COLOR 14:RETURN
2160 LOCATE 11,33:PRINT "Batch No.:";(NM+1)
2170 PRINT
2180 GOSUB 3960: REM Data To Stats Data Collection
     Subroutine
2190 GOTO 2080
2200 REM SUBROUTINE TO ASK IF PAUSE AFTER
     EACH TRIAL (Var. = PAU)
2210 CLS:LINE (500,110)–(150,220),2,B:PAINT
     (200,150),3,2
2220 LOCATE 10,23:PRINT "Do You Wish To Pause After
     Each Trial?"
2230 LOCATE 12,38:PRINT "(Y)es"
2240 LOCATE 13,38:PRINT "(N)o "
2250 LOCATE 15,30:PRINT B$;:P$=INPUT$(1)
2260 IF P$="Y" THEN PAU=1
2270 IF P$="y" THEN PAU=1
2280 IF P$="N" THEN PAU=2
2290 IF P$="n" THEN PAU=2
2300 IF PAU < 1 GOTO 2200
2310 IF PAU > 2 GOTO 2200
2320 IF INT(PAU) = PAU GOTO 2330 ELSE2200
2330 CLS
2340 RETURN
2350 REM SUBROUTINE FOR DONENESS
     COEFFICIENT (SUBJECTIVE OPINION)
```

```
2360  IF D>8 THEN GOTO 2540
2370  IF D < 2 THEN GOTO 2460
2380  IF D=2 THEN GOTO 2460
2390  IF D<4 THEN GOTO 2480
2400  IF D=4 THEN GOTO 2480
2410  IF D>4 THEN GOTO 2500
2420  IF D=6 THEN GOTO 2500
2430  IF D>6 THEN GOTO 2520
2440  IF D=8 THEN GOTO 2520
2450  GOTO 2560
2460  D$="Too Gooey To Use"
2470  GOTO 2450
2480  D$="Poorly Cured, Weak Joint"
2490  GOTO 2450
2500  D$="Done Properly, Strong Joint"
2510  GOTO 2450
2520  D$="Overcured, Joint Very Brittle and Fragile"
2530  GOTO 2450
2540  D$="Burned To a Crisp!"
2550  GOTO 2450
2560  RETURN
2570  REM PRESS ANY KEY SUBROUTINE
2580  PRINT "Press Any Key To Continue"
2590  A$=INKEY$:IF A$="" THEN 2590
2600  RETURN
2610  REM RANDOM NUMBER GENERATOR
2620  N=1000
2630  A%=VAL(MID$(TIME$,7,2))
2640  RANDOMIZE A%
2650  A=INT(RND*(N+1))
2660  IF A > 500 THEN 2670 ELSE2620
2670  IF A < 1000 THEN 2680 ELSE2620
2680  RETURN
2690  REM SUBROUTINE FOR NARROW NORMAL
      OVENS
```

```
2700 GOSUB 2610
2710 YY=A/2
2720 AA=SQR(A)
2730 DEBIAS=LEN(C$):IF DEBIAS > 5 THEN AA=AA
       ELSE AA=–AA
2740 AT=ST+(AA)
2750 AT =AT^2
2760 AT=SQR(AT)
2770 RETURN
2780 REM SUBROUTINE FOR WIDE NORMAL OVENS
2790 GOSUB 2610
2800 AA=A/10
2810 YY=A/2
2820 DEBIAS=LEN(C$):IF DEBIAS > 5 THEN AA=AA
       ELSE AA=–AA
2830 AT = ST + (AA)
2840 AT=AT^2
2850 AT=SQR(AT)
2860 RETURN
2870 REM SUBROUTINE FOR OUT OF CONTROL
       OVENS
2880 GOSUB 2610
2890 AT=A
2900 AT = A^2
2910 AT=SQR(AT)
2920 RETURN
2930 REM SUBROUTINE TO SELECT PROTOCOL FOR
       MULTIPLE TRIALS (Var.=PT)
2940 CLS:GOSUB 1040
2950 LOCATE 10,17:PRINT "Select Protocol For This Set of
       Trials       "
2960 LOCATE 12,17:PRINT "1. All Trials Same Time and
       Temperature       "
2970 LOCATE 13,17:PRINT "2. Same Temperature, Select
       New Time Each Trial"
```

```
2980  LOCATE 14,17:PRINT "3. Same Time, Select New
         Temperature Each Trial"
2990  LOCATE 15,17:PRINT "4. Return to Main Menu  "
3000  LOCATE 17,27:PRINT "Please Make Selection Now";
         :PT$=INPUT$(1)
3010  PT=VAL(PT$)
3020  IF PT<1 THEN GOTO 2930
3030  IF PT>4 GOTO 2930
3040  IF INT(PT)=PT GOTO 3050 ELSE2930
3050  CLS
3060  RETURN
3070  REM SUBROUTINE FOR NUMBER OF TRIALS
         (Variable = NT)
3080  CLS:GOSUB 940
3090  LOCATE 11,25:PRINT "Decide How Many Trials To
         Do"
3100  LOCATE 12,25:PRINT "(Allowable Range Is 2 to 50)"
3110  LOCATE 14,25:PRINT B$;:INPUT NT
3120  IF NT < 2 GOTO 3070
3130  IF NT > 50      GOTO 3070
3140  IF INT(NT)=NT GOTO 3150 ELSE3070
3150  CLS:GOSUB 1090
3160  RETURN
3170  REM RUN PROTOCOL PT=1 (Same Time/Temp
         Each Trial for NT Trials)
3180  NM=0:TBAR=0:DBAR=0
3190  CLS:GOSUB 1090
3200  GOSUB 2200
3210  GOSUB 1330:REM Select Oven (K)
3220  GOSUB 1880:REM Set Temperature (ST)
3230  GOSUB 1960:REM Set Time (T)
3240  ON K GOSUB 2690,2870,2690,2870,2780,2780:REM
         Get AT
3250  AT=CINT(AT)
3260  D = (AT/1000)*T:REM Calculate Doneness
```

3270 GOSUB 2350: REM Subjective Opinion Determination

3280 GOSUB 2050: REM Timer to reseed random number
     generator

3290 DBAR=DBAR+D:TBAR=TBAR+AT

3300 GOSUB 4360: REM Print Out Results of This Run

3310 NM=NM+1

3320 GOSUB 3960: DATA To Stats Data Collection
     Subroutine

3330 IF NM=NT GOTO 3350 ELSE3240

3340 IF NM=NT GOTO 3370 ELSE3240

3350 DBAR=DBAR/NT

3360 TBAR=TBAR/NT

3370 GOSUB 4020: REM Go Calculate Std. Dev. and
     Variance

3380 GOSUB 4590: REM Print Statistics For All Runs

3390 RETURN

3400 REM RUN PROTOCOL PT=2 (Set New Time Each
     Run/Same Temp)

3410 NM=0

3420 DBAR=0

3430 TBAR=0

3440 CLS:GOSUB 1090

3450 GOSUB 1330: REM Select Oven (K)

3460 GOSUB 1880: REM Set Temperature (ST)

3470 GOSUB 1960: REM Set Time (T)

3480 ON K GOSUB 2690,2870,2690,2870,2780,2780: REM
     Get AT

3490 AT=CINT(AT)

3500 D = (AT/1000)*T: REM Calculate Doneness

3510 GOSUB 2350: REM Subjective Opinion Determination

3520 GOSUB 2050: REM Timer To Reseed Random Number
     Generator

3530 DBAR=DBAR+D

3540 TBAR = TBAR + AT

3550 GOSUB 4360: REM Print Out Results of This Trial

```
3560  NM=NM+1
3570  GOSUB 3960: DATA To Stats Data Collection
         Subroutine
3580  IF NM=NT GOTO 3610 ELSE3650
3590  IF NM=NT GOTO 3630 ELSE3650
3600  PRINT TAB(5);"Actual Oven Temperature: ";AT
3610  DBAR=DBAR/NT
3620  TBAR=TBAR/NT
3630  GOSUB 4020: REM Go Calculate Std. Dev. and Variance
3640  GOTO 3670
3650  GOSUB 1960: REM Select New Time For Next Run
3660  GOTO 3480: REM Return to Beginning of Loop
3670  GOSUB 4590: REM Print Statistics For All Runs
3680  RETURN
3690  REM RUN PROTOCOL PT=3 (Same Time/New
         Temp. Each Trial).
3700  NM=0
3710  DBAR=0
3720  TBAR=0
3730  GOSUB 1330: REM Select Oven (K)
3740  GOSUB 1880: REM Set Temperature (ST)
3750  GOSUB 1960: REM Set Time (T)
3760  ON K GOSUB 2690,2870,2690,2870,2780,2780: REM
         Get AT
3770  AT=CINT(AT)
3780  D = (AT/1000)*T: REM Calculate Doneness
3790  GOSUB 2350: REM Subjective Opinion Determination
3800  GOSUB 2050: REM Timer To Reseed Random Number
         Generator
3810  DBAR=DBAR + D
3820  TBAR = TBAR + AT
3830  GOSUB 4360: REM Print Out Results of This Trial
3840  NM=NM+1
3850  GOSUB 3960: DATA To Stats Data Collection
         Subroutine
```

3860 IF NM=NT GOTO 3880 ELSE3920

3870 IF NM=NT GOTO 3900 ELSE3920

3880 DBAR=DBAR/NT

3890 TBAR=TBAR/NT

3900 GOSUB 4020: REM Go Calculate Std. Dev. and Variance

3910 GOTO 3940

3920 GOSUB 1880: REM Set New Oven Temperature (ST)

3930 GOTO 3760: REM Return to Beginning of Loop

3940 GOSUB 4590: REM Print Statistics For All Runs

3950 RETURN

3960 REM STATISTICS DATA COLLECTION
       SUBROUTINE

3970 NZ=NM+1

3980 DA(NZ)=D: REM Store Doneness Coefficient For This
       Iteration

3990 AX(NZ)=AT: REM Store Actual Temperature For This
       Iteration

4000 SB(NZ)=ST: REM Store Set Temperature For This
       Iteration

4010 RETURN

4020 REM SUBROUTINE TO CALCULATE MEAN,
       STD.DEV., VAR.

4030 V=0:W=0:VARD=0:VART=0:DEVT=0:DEVD=0:
       WX=0:V=0:VX=0

4040 CLS:GOSUB 940:LOCATE 12,22:PRINT "Making
       Calculations...Please Wait"

4050 REM Calculate Variances

4060 FOR II = 1 TO NT

4070 V = (DA(II)–DBAR)^2

4080 W = (AX(II)–TBAR)^2

4090 WX = WX + W

4100 VX = VX + V

4110 NEXT II

4120 VARD = VX/(NT)

4130 VART = WX/(NT)

```
4140 DEVD=SQR(VARD)
4150 DEVT=SQR(VART)
4160 REM SUBROUTINE TO SORT TEMPERATURE/
        DONENESS RAW DATA POINTS
4170 FOR II = 1 TO NT: REM Preset arrays
4180 MINT(II) = AX(II)
4190 MIND(II) = DA(II)
4200 NEXT II
4210 FOR JJ = 1 TO NT - 1
4220 FOR KK = JJ + 1 TO NT
4230 IF MIND(KK) < MIND(JJ) THEN SWAP
        MIND(KK),MIND(JJ)
4240 NEXT KK
4250 NEXT JJ
4260 FOR JJ = 1 TO NT - 1: REM Sort Temperature Data
4270 FOR KK = JJ + 1 TO NT
4280 IF MINT(KK) < MINT(JJ) THEN SWAP
        MINT(KK),MINT(JJ)
4290 NEXT KK
4300 NEXT JJ
4310 MINTEMP = MINT(1)
4320 MAXTEMP = MINT(NT)
4330 MAXDONE = MIND(NT)
4340 MINDONE = MIND(1)
4350 RETURN
4360 REM PRINT OUT RESULTS FOR MULTIPLE
        TRIALS IF SAME TEMP/TIME
4370 CLS:GOSUB 1040
4380 AT=100*AT
4390 AT=INT(AT)
4400 AT=AT/100
4410 D=100*D
4420 D=INT(D)
4430 D=D/100
4440 LOCATE 7,30:PRINT "RESULTS FOR THIS TRIAL"
```

```
4450 LOCATE 9,22:PRINT "Trial No. ";NM+1
4460 LOCATE 10,22:PRINT "Set Temperature: ";ST;"
        Degrees F."
4470 LOCATE 11,22:PRINT "Actual Temperature: ";AT;"
        Degrees F.
4480 LOCATE 13,22:PRINT "Curing Time Selected:";T
4490 LOCATE 14,22:PRINT "Doneness (Zlotz Units): ";
4500 PRINT USING "##.##";D
4510 LOCATE 16,22:PRINT "Subjective Opinion of
        Doneness: "
4520 LOCATE 17,22:PRINT D$
4530 IF PAU = 2 GOTO 4570 ELSE4540
4540 PRINT
4550 PRINT
4560 LOCATE 20,27:GOSUB 2570: REM Press Any Key...
4570 TIMELOOP=TIMER:WHILE
        TIMER<TIMELOOP+4:WEND
4580 RETURN
4590 REM STATISTICS PRINTOUT SUBROUTINE
4600 GOSUB 5460: REM Round off values for printing
4610 CLS:GOSUB 1090:COLOR 7,1:COLOR 14
4620 PRINT TAB(20);"STATISTICS FOR ALL TRIALS
        IN THIS SERIES"
4630 PRINT
4640 PRINT TAB(20);"TEMPERATURE"
4650 PRINT TAB(20);"Mean Temperature: ";TBAR;"
        Degrees F."
4660 PRINT TAB(20);"Variance: ";
4670 PRINT USING "########.#####";VART
4680 PRINT TAB(20);"Standard Deviation: ";
4690 PRINT USING "#######.#";DEVT
4700 PRINT
4710 PRINT TAB(20);"DONENESS"
4720 PRINT TAB(20);"Mean Doneness: ";
4730 PRINT USING "##.#########";DBAR
```

```
4740  PRINT TAB(20);"Variance: ";
4750  PRINT USING "##.######";VARD
4760  PRINT TAB(20);"Standard Deviation: ";
4770  PRINT USING "##.#######";DEVD
4780  PRINT
4790  PRINT TAB(20);"RANGES:"
4800  PRINT TAB(20);"Temperature: ";MINTEMP;" To
       ";MAXTEMP;" Degrees F."
4810  PRINT TAB(20);"Doneness: ";MINDONE;" To
       ";MAXDONE;" Zlotz Units."
4820  PRINT
4830  PRINT
4840  PRINT TAB(20):GOSUB 2570
4850  CLS: GOSUB 990
4860  LOCATE 10,23:PRINT "Would You Like a Hard Copy
       Output?"
4870  LOCATE 12,36:PRINT "(N)o "
4880  LOCATE 13,36:PRINT "(Y)es"
4890  LOCATE 15,29:PRINT B$;:LNPRT$=INPUT$(1)
4900  IF LNPRT$="N" THEN LNPRT=1
4910  IF LNPRT$="n" THEN LNPRT=1
4920  IF LNPRT$="Y" THEN LNPRT=2
4930  IF LNPRT$="y" THEN LNPRT=2
4940  IF LNPRT < 1 GOTO 4850
4950  IF LNPRT > 2 GOTO 4850
4960  IF INT(LNPRT) = LNPRT GOTO 4970 ELSE4850
4970  CLS
4980  ON LNPRT GOTO 5450,4990
4990  CLS:GOSUB 1090:PRINT TAB(35);"PRINTING..."
5000  ON LNPRT GOTO 5450,5010
5010  LPRINT TAB(15);"OVEN NUMBER: ";K
5020  LPRINT:LPRINT TAB(15);"Experimenter: ";C$
5030  ON PT GOSUB 5330,5370,5410
5040  LPRINT
5050  FOR II = 1 TO NT
```

```
5060 LPRINT TAB(15);"Run";II;" Actual Temp.: ";AX(II);"
        Doneness: ";
5070 LPRINT USING "##.##";DA(II)
5080 NEXT II
5090 LPRINT
5100 LPRINT
5110 LPRINT TAB(15);"TEMPERATURE"
5120 LPRINT TAB(15);"Mean Temperature: ";TBAR;"
        Degrees F."
5130 LPRINT TAB(15);"Variance: ";
5140 LPRINT USING "########.";VART
5150 LPRINT TAB(15);"Standard Deviation: ";
5160 LPRINT USING "#######.##";DEVT
5170 LPRINT
5180 LPRINT TAB(15);"DONENESS"
5190 LPRINT TAB(15);"Mean Doneness: ";
5200 LPRINT USING "##.##";DBAR
5210 LPRINT TAB(15);"Variance: ";
5220 LPRINT USING "##.#######";VARD
5230 LPRINT TAB(15);"Standard Deviation: ";
5240 LPRINT USING "##.#######";DEVD
5250 LPRINT
5260 LPRINT TAB(15);"RANGES:"
5270 LPRINT TAB(15);"Temperature: ";MINTEMP;" To
        ";MAXTEMP;" Degrees F."
5280 LPRINT TAB(15);"Doneness: ";MINDONE;" To
        ";MAXDONE;" Zlotz Units."
5290 LPRINT
5300 LPRINT
5310 BEEP:PRINT:PRINT TAB(28):GOSUB 2570:
5320 GOTO 5450
5330 REM Print routine for same Time/Temp each trial
        (PT=1)
5340 LPRINT TAB(15);"Set Oven Temperature: ";ST
5350 LPRINT TAB(15);"Selected Curing Time: ";T
```

```
5360 RETURN
5370 REM Print routine for same temp./new time each trial
        (PT=2)
5380 LPRINT TAB(15);"Set Oven Temperature: ";ST
5390 LPRINT TAB(15);"Selected Curing Time: ";G$
5400 RETURN
5410 REM Print routine for Temp.New/Time same each trial
        (PT=3)
5420 LPRINT TAB(15);"Set Oven Temperature: ";H$
5430 LPRINT TAB(15);"Selected Curing Time: ";T
5440 RETURN
5450 RETURN
5460 RETURN:REM SUBROUTINE DELETED FOR TEST
        PURPOSES
5470 TBAR = CINT(TBAR)
5480 DEVT = CINT(DEVT)
5490 MINTEMP = CINT(MINTEMP)
5500 MAXTEMP = CINT(MAXTEMP)
5510 AT = CINT(AT)
5520 D = D*100
5530 D = CINT(D)
5540 D = D/100
5550 MINDONE = MINDONE*100
5560 MINDONE = CINT(MINDONE)
5570 MINDONE = MINDONE/100
5580 MAXDONE = MAXDONE*100
5590 MAXDONE = CINT(MAXDONE)
5600 MAXDONE = MAXDONE/100
5610 RETURN
5620 REM Subroutine to print time remaining on screen
5630 TT = 10*T:XT=T:MT=T
5640 LOCATE 17,30:PRINT "Time remaining:"
5650 LOCATE 17,46:PRINT USING "###.#";XT;
5660 TIMELOOP1=TIMER:WHILE TIMER < TIMELOOP1
        + (XT/4):WEND
```

```
5670  IF XT < .1 THEN 5710 ELSE5680
5680  REM
5690  IF XT < 0 THEN 5710 ELSE5700
5700  XT=XT-.1:GOTO 5650
5710  RETURN
```

Index